Towards Unmanned Surface Vehicles

Towards Unmanned Surface Vehicles: Methods and Practices presents the latest overview, methodologies, design practices, and applications of unmanned surface vehicles (USVs).

The authors introduce advanced theories and algorithms for the analysis and design of a maritime unmanned surface vehicle system, covering the sensing, path following, navigation, and control of the ocean surface environment. They demonstrate the architectural design, implementation, and field testing of USVs as well as key applications, such as hostile military scenarios, scientific oceanographic observation, and intelligent waterborne transportation. In addition, they address the open challenges in the field and propose the corresponding future perspectives.

The book will appeal to researchers, graduate students, and engineers interested in USVs.

Huarong Zheng is a full researcher at State Key Laboratory of Ocean Sensing & Ocean College, Zhejiang University, Zhoushan, China. She has been working on autonomous maritime vehicles for more than ten years. Her research interests include ocean robotics, model predictive control and distributed control and coordination with applications to waterborne networked systems.

Chenguang Liu is a professor at State Key Laboratory of Maritime Technology and Safety, Wuhan University of Technology, China. His research interests include ship intelligent navigation, motion planning and control, formation control, advanced perception, and model predictive control theories and applications.

Towards Unmanned Surface Vehicles

Methods and Practices

Huarong Zheng and Chenguang Liu

CRC CRC Press
Taylor & Francis Group
Boca Raton London New York

CRC Press is an imprint of the
Taylor & Francis Group, an **informa** business

Designed cover image: © Chenguang Liu

This book is published with financial support from National Natural Science Foundation of China under Grants 62473332 and 52001240

MATLAB® and Simulink® are trademarks of The MathWorks, Inc. and are used with permission. The MathWorks does not warrant the accuracy of the text or exercises in this book. This book's use or discussion of MATLAB® or Simulink® software or related products does not constitute endorsement or sponsorship by The MathWorks of a particular pedagogical approach or particular use of the MATLAB® and Simulink® software.

First edition published 2026
by CRC Press
2385 NW Executive Center Drive, Suite 320, Boca Raton, FL 33431

and by CRC Press
4 Park Square, Milton Park, Abingdon, Oxon, OX14 4RN

CRC Press is an imprint of Taylor & Francis Group, LLC

ISBN: 978-1-041-11559-5 (hbk)
ISBN: 978-1-041-11560-1 (pbk)
ISBN: 978-1-003-66059-0 (ebk)

DOI: 10.1201/9781003660590

Typeset in LMRoman
by KnowledgeWorks Global Ltd.

Contents

SECTION III Applications over the Water

CHAPTER   9 ▪ Waterborne Applications of USVs   175

CHAPTER   10 ▪ Summary and Future Perspectives   182

APPENDIX   A ▪ Appendix   191

Preface

Unmanned Surface Vehicles (USVs) are a special type of unmanned vehicles. Many technologies of USVs can inherit from unmanned vehicles in general. However, USVs face particular challenges due to the complex maritime environment. Gaining intelligence in the uncertain, unstructured, and resource-limited maritime environment calls for extra efforts. Moreover, the technology tests, practices, and operations in the remote, dangerous, and harsh maritime environment are also more demanding both technically and economically.

Both authors of the book have been working on improving the intelligence and promoting the practice of USVs during the past ten years. The past decade has seen that maritime USVs briskly move from the concept to reality. The rapid development of USVs is driven by both the progressing technologies and the emerging scientific, engineering, and military demands. The focus of USVs has shifted from pure research activities to commercial and engineering applications. To perform satisfactorily in different applications and tasks, the fundamental requirements for USVs are the agility in dynamic and complex maritime conditions. Agility of USVs builds on the advanced autonomous navigation, guidance, and control. Achieving agility is not easy in the complex maritime environment. The environment perception sensors such as cameras, lidars, and radars in the navigation module might fail due to the water surface reflection or mist. The irregular waterways and complicated encountering scenarios necessitate careful design of the USV guidance module. Challenges also come from the control module in that real-time control of USVs with uncertain hydrodynamics, large inertia, and weak maneuverability is especially difficult.

This book describes the insights gained by the authors from a decade of USV development, and presents the latest overview, methodologies, design practices, and applications of USVs. The target readers of the book are the researchers, graduate students, and engineers that are interested in USVs. The goal in writing this book is threefold. Firstly, the book provides a systematic update on the advanced methodologies of USVs. Secondly, the book covers the relatively complete topics in the design of USVs. Readers can learn broadly about USVs or read a specific chapter for details. Finally, we try to connect methodologies, algorithms, and practices of USVs in this book. In this way, readers can gain the knowledge of advanced methods, and at the same time know how to implement these methods.

This book is divided into three parts. Prior to Part I, Chapter 1 first introduces the background of USVs. Typical USVs to date, including those both for research purposes and for field practices, are comprehensively reviewed and compared. Then, Part I Advanced Navigation, Guidance and Control Methods Towards Autonomy (Chapters 2-5) focuses on the state-of-the-art methodologies that constitute the core capabilities of USVs. The emerging data-driven and learning-based approaches are also elaborated. Part II Practices in the Field (Chapters 6-8) reflects on the work of the architectural design, implementation, and field tests of USVs. The last part (Chapter 9) introduces the major applications of USVs, e.g., oceanographic observation and intelligent waterborne transportation. Conclusions and future perspectives of the authors are also given at the end of the book in Chapter 10.

Most contents of this book refer to the authors' research work at Zhejiang University, Wuhan University of Technology, and Delft University of Technology during the last ten years. A large number of people have been working together for conducting the research in this book. The authors' supervisors and colleagues, Professor Xinping Yan, Professor Wen Xu, Professor Rudy R. Negenborn, Professor Chaozhong Wu, and Professor Xiumin Chu, are highly appreciated for their outstanding guidance in the authors' early academic career. Also, we would like to thank our students: Jiacheng Li, Yan Zhou, Puling Zhang, Jiangbo Zhu, Wenxiang Wu, Zixiang Jiang, and Zhibo He for conducting solid works on USVs in the research groups. Being supervised and now supervising, the authors firmly stand the role of carrying forward both knowledge and spirit as researchers and educators.

Lastly, Zixu Fan has provided excellent support throughout the final stages of preparation of this book. We would also like to thank our commissioning editor, Lian Sun, for her support and professionalism.

Huarong Zheng
Zhoushan, China

Chenguang Liu
Wuhan, China

Glossary

a	binary vector parameterizing the distribution function
A	Jacobian state matrix
$A_{K,\rho}$	state matrix with feedback
b	environmental disturbances
$\bar{b}$	predicted disturbance force
B	Jacobian control input matrix
C_u	convex set constraints on perturbation control inputs of USV
C_x	convex set constraints on perturbation states of USV
d_s	safety distance from the obstacle
D_s	minimum safety distance between USVs
E	Jacobian disturbance matrix
I_x, I_u	number of state and control input constraints
k	discrete time step
K	time-varying feedback gain
$\{n\}$	the inertial coordinate system
N_p	prediction horizon
p	probability of $b \in [-z, z]$
P	discrete probability vector
$\mathcal{P}$	convex sets of uncertain perturbation states of USV
$\Delta \bar{P}$	nominal perturbation states of USV
Q_f	state cost weight matrix
$\mathcal{Q}_f$	terminal state cost weight matrix
r	yaw rate of USV
$\mathcal{R}$	control input weight matrix
T	diagonal translation matrix

T_s	sampling time
t	continuous time
u_{min}, u_{max}	minimum and maximum speeds of USV
$\tilde{u}$	uncertain perturbation control inputs of USV
Δu	nominal perturbation control inputs of USV
$\boldsymbol{U}$	system control input vector of USV
$\mathcal{U}$	control input tube of USV
v	sway force of USV
$v_{\min}$	minimum velocity vector of USV
$v_{\max}$	maximum velocity vector of USV
v_r	relative velocity vector of USV with respect to current speed
$\boldsymbol{V}$	velocity vector of USV
V_c	current speed
x	coordinate along X_n axis of USV
$\boldsymbol{x}$	system state vector of USV
$\Delta \boldsymbol{x}$	uncertain perturbation states of USV
$\mathcal{X}$	state tube of USV
y	coordinate along Y_n axis of USV
z	uncertainty bounds
$\mathcal{Z}$	discrete bound vector
β_c	current angle
η	pose vector of USV
ψ	heading angle of USV
τ	control input vector of USV
$\tau_{\min}$	minimum control input vector of USV
$\tau_{\max}$	maximum control input vector of USV
τ_r	yaw moment of USV
τ_u	surge force of USV
τ_v	sway force of USV

Acronyms

waterborne AGVs	waterborne Autonomous Guided Vessels
ACO	Ant Colony Optimization
AIS	Automatic Identification System
APF	Artificial Potential Field
ARPA	Automatic Radar Plotting Aid
ASVs	Autonomous Surface Vehicles
AUVs	Autonomous Underwater Vehicles
CB	Constant Bearing
CFD	Computational Fluid Dynamics
CMVSs	Cooperative Multi-Vessel Systems
COLREGs	International Regulations for Preventing Collisions at Sea
CPA	Closest Point of Approach
CPR	Center Point Restoration
CRI	Collision Risk Indices
CSFSs	Cooperative Ship Formation Systems
DARPA	Defense Advanced Research Projects Agency
DCPA	Distance to the Closest Point of Approach
DNN	Deep Neural Network
DMPC	Distributed Model Predictive Control
DOF	Degree-Of-Freedom
ENC	Electronic Navigation Chart
GA	Genetic Algorithm
GNSS	Global Navigation Satellite System
GP	Gaussian Process
GPS	Global Positioning System

HMI	Human-Machine Interaction
IMU	Inertial Measurement Unit
JPDA	Joint Probability Data Association
K-NN	k-Nearest Neighbor
LCD	Liquid Crystal Display
LIDAR	Light Detection and Ranging
LOS	Line Of Sight
LQG	Linear Quadratic Gaussian
LSTM	Long Short-Term Memory
MASS	Maritime Autonomous Surface Ships
MMG	Mathematical Model Group
MMW	Milli Meter-Wave
MOOS	Mission Oriented Operating System
MPC	Model Predictive Control
MSS	Marine Systems Simulator
NGC	Navigation, Guidance and Control
NTNU	Norwegian University of Science and Technology
PCL	Point Cloud Library
PID	Proportional-Integral-Derivative
PSO	Particle Swarm Optimization
RCS	Radar Cross Section
RL	Reinforcement Learning
ROS	Robot Operation System
RRT	Rapidly-exploring Random Tree
SLAM	Simultaneous Localization and Mapping
TCPA	Time to Closest Point of Approach
UAVs	Unmanned Aerial Vehicles
UMVs	Unmanned Marine Vehicles
UWB	Ultra Wide Band
USVs	Unmanned Surface Vehicles
VDE	Velocity Distribution Extraction

CHAPTER 1

Overview of Unmanned Surface Vehicles

1.1 INTRODUCTION

A UTONOMOUS mobile robots, particularly Unmanned Surface Vehicles (USVs), have rapidly advanced in recent years, benefiting from multi-disciplinary robotics technologies. These vehicles, which vary in application and are known by several names including Autonomous Surface Vehicles (ASVs) [1], Unmanned Marine Vehicles (UMVs) [3], waterborne Autonomous Guided Vessels (water AGVs) [4][5], and Maritime Autonomous Surface Ships (MASS) [6], represent a diverse class of ships that can operate with varying degrees of independence from human operators [7]. This book will explore theories and practices pertinent to all these terms, using the designation USVs consistently throughout.

As technological advancements continue to mature, USVs are increasingly essential in waterborne engineering, scientific research, and military applications. They offer significant advantages over traditional crewed surface vehicles in terms of safety, comfort, flexibility, and efficiency. USVs are viewed as a transformative force that could revolutionize the maritime sector [8]. Currently, various countries and organizations have the opportunity to lead in this emerging field, with European and Asian nations, as well as the United States, particularly active in enhancing competitiveness. Notably, the Norwegian company Maritime Robotics has gained approval to establish the world's first freight route for uncrewed cargo boats [9]. Additionally, the Chinese company Yunzhou has successfully demonstrated the operation of 56 USVs in swarm formations at sea [10]. Dutch researchers have recently highlighted in Nature that this is an ideal time for the industry to concentrate on USVs [8].

1.2 USV RELEVANT ACTIVITIES AND EFFORTS

Naturally, numerous key projects, non-profit collaborations, and strategic roadmaps for USVs have been established over the past decade, as summarized in Table 1.1. Generally, European and Asian countries have primarily focused on the civilian applications of USVs, such as autonomous shipping. In contrast, the United States

DOI: 10.1201/9781003660590-1

TABLE 1.1 USV relevant major strategic activities in the recent 10 years.

Region	Year	Activities	Brief introduction
EU	2017	H2020 project: NOVIMAR [12]	Waterborne transport system with vessel trains
	2019	H2020 project: AUTOSHIP [13]	Speeding-up the transition towards the next generation of autonomous ships
	2020	H2020 project: MOSES [14]	Sustainable short-sea shipping with automated vessels and supply chain optimization
	2020	H2020 project: AEGIS [15]	Waterborne logistics system with autonomous ships and automated ports
US	2020	DARPA project: No Manning Required Ship [16]	Ship design without crew and supports distributed maritime operations
	2020	DARPA project: Sea Train [17]	Enhancing ranging with connected vessels
	2021	Navy project: Unmanned Campaign Framework [18]	Roadmap to scale tested and proven unmanned systems to fleet
Norway	2022	National project: SAFEMATE [19]	Evaluate the efficiency and safety of autonomous navigation systems
Netherlands	2021	Alliance: SMASH! [20]	Netherlands forum for smart shipping
Denmark	2019	Institute: ShippingLab [21]	Create Denmark's first autonomous, environmental friendly ship
Finland	2016	Alliance: One Sea [22]	Global alliance for an automated and autonomous maritime transport system
Japan	2020	Institute: MEGURI 2040 [23]	Implement fully autonomous navigation
South Korea	2020	National project: KASS [24]	Korea Autonomous Surface Ship Project
UK	2019	Roadmap: Maritime 2050: Navigating the Future [25]	Autonomous Navigation Innovation Hub by 2025
Singapore	2019	Institute: Maritime Innovation Lab (MIL) [26]	Support new port services and intelligent ship operations

excels in technological development by supporting a wide range of military projects, although it lags behind European nations in the autonomous shipping sector [11]. Several notable USV competitions have emerged, including the Maritime RobotX Challenge, the World Robotics Sailing Championship, the Microtransat Challenge, the Shell Ocean Discovery XPRIZE, the Zhuhai Wanshan International Intelligent Vessel Competition, the National Ocean Vehicles Competition, and initiatives led by the US Defense Advanced Research Projects Agency (DARPA).

Significant efforts from both academia and industry have been devoted to advancing the methods, practices, and applications of USVs. In recent years, there has been considerable research progress in this area. Various review articles [27, 2, 28, 29, 30, 31, 32, 33, 34] have been published to update developments and outline potential challenges associated with USVs. Some of these reviews focus on specific subfields, such as collision avoidance [27], overactuated USVs [28], safety perspectives [29], coordinated control [30], and deep learning applications related to USVs [33].

Additionally, some reviews discuss future considerations for USVs transitioning from research concepts to commercial products [31] and their application in inland shipping [32]. Notably, [2] presents a comprehensive review that organizes approximately 200 papers on USVs up to 2016. Nearly a decade later, this review has been cited almost 900 times, primarily by works related to USVs. The high citation count not only reflects significant advancements in the USV field since 2016 but also highlights the growing interest and popularity of this topic. While existing review works predominantly emphasize the algorithmic and methodological aspects of USVs, notable recent advancements highlight the successful applications of various commercial USV products [9, 10, 34]. Over the past several decades, research and development have shifted focus toward practical validations and application-oriented capabilities built upon methodological advances.

Practical validation of USVs involves careful consideration of power and data flows among different subsystems, integrating mechanical design, software development, circuit modules, payload selection, and USV-to-shore communications. The complex marine environment presents unique challenges to the reliability of USV systems and the validation testing procedures. However, the ability to operate autonomously in uncertain and unstructured marine environments without human intervention is invaluable for many applications. Despite these advancements, application-oriented capabilities in specific engineering, scientific, and military scenarios require further development. The future progress of USVs will depend on a holistic approach that integrates the development of methods, practices, and application capabilities.

Therefore, this book provides the state-of-the-art on the latest USV advances with respect to the methodologies, design practices and applications. With the insights gained by the authors working in the field for more than a decade, we envision the three aspects, i.e., methods towards autonomy, practices in the field, and application capabilities, are of the uttermost and equivalent importance in the further progress of USVs. Correspondingly, after introducing the general aspects of USVs in Chapter 1, the book is organized in three parts as:

- **Part I**: The state-of-the-art Navigation, Guidance and Control (NGC) methods are presented, including both the classical techniques that constitute the fundamental autonomy of USVs and the emerging learning-based ever-powerful technologies. Specifically, the typical USV navigation methods are discussed in Chapter 2. Chapter 3 introduces the USV guidance and planning problems, followed by the USV control problems in Chapter 4. Chapter 5 further extends the discussions to the emerging data-driven and learning-based USV control approaches.

- **Part II**: The USV systematic design, implementation, and test practices are presented. Chapter 6 reflects on the work of the USV architectural design. Two field tests, lock waterway navigation and turning maneuver tests, are detailed in Chapter 7 and Chapter 8, respectively.

- **Part III**: The application capabilities of USVs in different fields, e.g., military, intelligent waterborne transportation, and oceanographic observation are discussed in Chapter 9.

Conclusions and future perspectives of the authors are given at the end of the book in Chapter 10. Most chapters are accompanied by relevant state-of-the-art USV technology examples and algorithm details. The road map of this book is briefly presented in Fig. 1.1, illustrating connections of chapters and a suggested order in which the chapters can be read.

To the best of the authors' knowledge, this is the first book that comprehensively summarizes the relevant methods, practices, and applications of USVs. By detailing state-of-the-art advancements, this book aims to support both the USV research and industry communities in gaining a clear overview of the field, enhancing research efficiency, and accelerating the development of practical USV applications.

1.3 STATE-OF-THE-ART USV PLATFORMS

A decade ago, most work on USVs was primarily focused on research, with small-scale prototypes developed and tested in laboratory settings [35, 36, 4]. As methodologies and technologies have matured, we now observe an increase in USVs featuring higher levels of automation, which are being applied in diverse fields. This section provides an update on the current USV platforms available.

1.3.1 Automation levels

Before proceeding to various USV platforms, we first clarify the definitions of different automation levels in USVs. The core automation capabilities of USVs depend on their sensors, computing processors, and algorithms. As technology advances in industrial sensors, processing units, and algorithms, intelligent ships equipped with automated co-pilots or autopilots require less human supervision and intervention during voyages.

The International Maritime Organization (IMO) has distinguished four degrees of autonomy for USVs based on varying levels of human-machine interaction [7]. While other classifications of autonomy levels exist, such as those by the Norwegian Classification Society [37], interested readers can refer to [38] for a detailed survey on USV autonomy levels. This discussion will focus on the IMO definitions due to their widespread adoption.

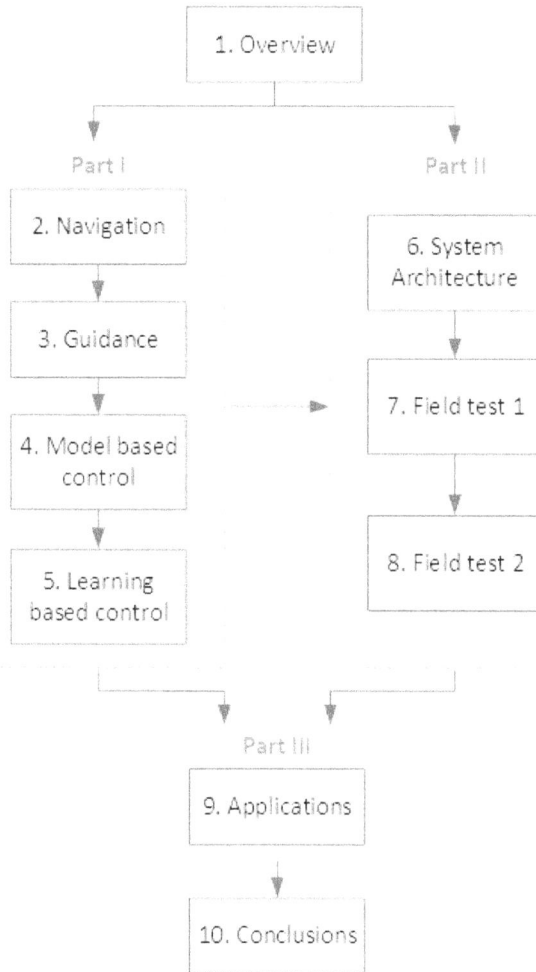

Figure 1.1: Roadmap of the book. Arrows indicate *read before* relations.

Compared to the common taxonomy for automated driving levels adopted by the US transportation administration [39], the IMO's USV automation levels place greater emphasis on remote operation modes. Table 1.2 presents a brief comparison between the automation levels defined for intelligent cars and intelligent ships. For both vehicle types, the lowest three automation levels still require human monitoring of the environment. However, the shipping industry accepts a level of remote control without seafarers on board, which is classified below full automation [40]. This acceptance may stem from the fact that ships are primarily used for transporting cargo, coupled with the generally harsh working conditions at sea. Nonetheless, it is important to note that the remote-control mode still necessitates human attention and interaction. Challenges related to remote control, such as communication and security issues [41], must also be addressed.

With the highest level of full automation, zero human attention or interaction is required. The higher automation level directly reduces the workload of seafarers, giving them more social time with friends and families. The human errors that are the

TABLE 1.2   Comparison on the intelligence levels of intelligent cars and intelligent ships [39][7].

	The human monitors the environment			The machine monitors the environment		
Intelligent cars	*Level 0*	*Level 1*	*Level 2*	*Level 3*	*Level 4*	*Level 5*
	No automation	Driver assistance	Partial automation	Conditional automation	High automation	Full automation
Intelligent ships	*Level 1*	*Level 2*	*Level 3*	*Level 4*	/	/
	Ship with automated processes and decision support	Remotely controlled with seafarers on board	Remotely controlled without seafarers on board	Full automation	/	/

most important factor contributing to maritime accidents [42] could also be reduced. Particularly, USVs with full automation see the benefits at least on the following aspects:

(1) Reducing the workload of seafarers;

(2) Reducing human-related accidents;

(3) Reducing the personnel costs;

(4) More efficient and environmentally friendly navigating behaviors;

(5) Long operational times in 24/7;

(6) Lowering the possible casualties in dangerous situations;

(7) In line with the development of smart industries, e.g., intelligent transportation system.

The ultimate goal is to achieve reliable Level 4 full automation for USVs without human interaction. While significant progress has been made in remote control capabilities, as seen in studies such as [43] and [44], this book will focus on discussing Level 4 USV platforms and their developments. Regardless of their size, appearance, or application, the core of Level 4 USVs lies in their intelligence. This intelligence is generally realized by sensing the environment, processing the collected data, and determining actions to accomplish assigned missions. For USVs, these processes are facilitated by NGC modules and algorithms [28]. Fig. 1.2 provides a brief overview of the technological components of such an NGC system. The NGC modules predominantly rely on theoretical tools for estimation, optimization, and control, along with emerging learning-powered techniques, which will be discussed in greater detail in subsequent chapters.

1.3.2   An update on the major USV platforms

The past decade has witnessed rapid development in USVs. We can clearly observe a significant transformation from small-scale experimental platforms to fully

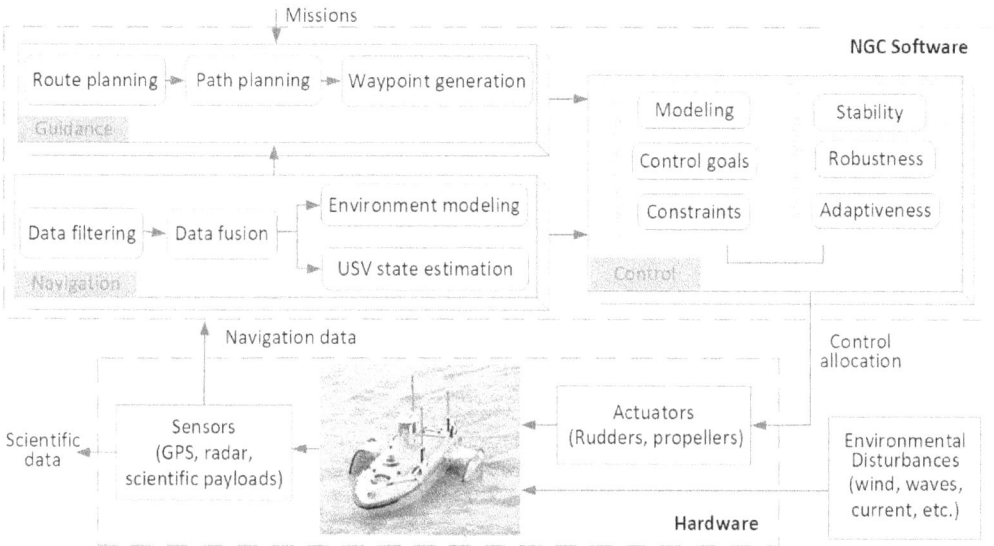

Figure 1.2: The NGC system of USVs.

operational intelligent ships, as well as a shift from pure laboratory research to field tests and practical applications.

Research on USVs gained momentum at the beginning of the 21st century, leading to the development of a wide variety of prototypes, primarily driven by institutions and universities. Notable contributions include a series of unmanned vessels designed by the Massachusetts Institute of Technology (MIT), such as ARTEMIS (1993), ACES (1997), AutoCat (1999), and SCOUT (2004) [45]. MIT also developed the Mission Oriented Operating System (MOOS) [46] for their unmanned marine crafts, encompassing both surface and underwater vehicles. The University of Genova in Italy produced platforms like ALANIS and Charlie [47], which have applications in mine hunting and Arctic/Antarctic operations. In Portugal, the Institute for Systems and Robotics developed the Delfim series (Delfim, Caravelas, and DelfimX) [48] for cooperation with unmanned submersibles and data acquisition. The Cybership series [49] from the Norwegian University of Science and Technology (NTNU) has gained recognition in academia for providing public system information and the Marine Systems Simulator (MSS) software toolkit [50]. Other noteworthy scientific platforms include the Measuring Dolphin [51] from the University of Rostock, Germany; Springer [52] from Queen's University Belfast, UK; Sillverlit from Eötvös Loránd University, Hungary [53]; and Delfia [54] from Delft University of Technology, Netherlands. For more comprehensive surveys on these early USV prototypes, interested readers are referred to [4], [2], and [55] for additional details.

In the past decade, laboratory prototypes have transitioned into mature commercial and non-commercial products. As a result, business companies and investors have begun to engage in the field of USVs. Traditional ship types, including container ships, ferries, sailboats, tugboats, and survey vessels, have all experienced advancements in autonomy. Table 1.3 provides an update on the major USV platforms and

TABLE 1.3 Major USV platforms in the recent 10 years.

Country	Year	Organization	Platform	Application
Norway	2023	Maritime Robotics	Mariner [9]	Uncrewed freight transport at sea
	2022	NTNU	MilliAmpere 2 [56]	Urban autonomous passenger ferry; Predecessor *milliAmpere* in 2016
	2021	Yara & Kongsberg	Yara Birkeland [57]	World's first fully electric and autonomous container ship with zero emissions
	2014	DNV & NTNU	ReVolt [58]	Test the next-generation unmanned shipping concept
Finland	2018	Finferries & Rolls-Royce	Falco [59]	World's first fully autonomous ferry
	2020	MIT	Roboat II [60]	Transport over urban waterways
US	2018	SailDrone	SailDrone USV [61]	Wind-propelled long-range ocean data collection
	2016	DARPA	Sea Hunter [62]	Anti-submarine warfare
China	2018	Huazhong University of Science and Technology	HUSTER USVs [63]	For survey, patrol, surveillance, and law enforcement missions
	2017	Shanghai University	Jinghai USVs [64]	Hydrographic survey and mapping in shallow waters
	2013	Yunzhou Intelligence Technology	Yunzhou USVs [65]	Oceanographic survey and seafloor mapping
Netherlands	2019	Delft University of Technology	Grey Seabax [66]	Small scaled offshore support vessel
Japan	2022	Imoto Corporation	Mikage [23]	Autonomous container transport
South Korea	2022	Hyundai	Prism Courage [67]	Large-size LNG transport ship
Croatia	2017	H2O Robotics	H2Omni-X [68]	Dynamic positioning and diver following
UK	2019	AutoNaut	AutoNaut USV [69]	Wave-propelled uncrewed surface vessels for long-range ocean observation
	2017	SEA-KIT	SEA-KIT USV [70]	Autonomous survey or defense vessel with adaptive payload area
Singapore	2020	Wärtsilä & PSA Marine	PSA Polaris [71]	Autonomous harbor tugboat

Figure 1.3: Different product-level USV platforms for different applications. (a) The world's first autonomous container ship, Yara Birkeland [57]. (b) The world's first fully autonomous ferry [59]. (c) H2Omni-X [68] Modular unmanned surface vehicle by H2O Robotics. (d) Wind-propelled long-range ocean data collection USV Saildrone [61].

products developed in the last ten years. Additionally, Fig. 1.3 illustrates four representative types of product-level USV platforms, as listed in Table 1.3, showcasing their diverse applications.

1.4 CONCLUSIONS

This chapter provides an overview of the development of USVs over the past decade. Research has evolved from military and purely academic purposes to commercial and practical applications. Notably, the chapter comprehensively organizes major strategic activities related to USVs and highlights key product-level platforms from the last ten years. Additionally, it compares and discusses two types of intelligent vehicles: unmanned cars and unmanned marine surface vehicles. With this overview of USVs established, the remainder of the book will introduce specific advanced USV technologies and their implementations.

Bibliography

[1] L. Hu, W. Naeem, E. Rajabally, G. Watson, T. Mills, Z. Bhuiyan, C. Raeburn, I. Salter, and C. Pekcan, "A multiobjective optimization approach for COLREGs-compliant path planning of autonomous surface vehicles verified on networked bridge simulators," *IEEE Transactions on Intelligent Transportation Systems*, vol. 21, no. 3, pp. 1167–1179, 2020.

[2] Z. Liu, Y. Zhang, X. Yu, and C. Yuan, "Unmanned surface vehicles: An overview of developments and challenges," *Annual Reviews in Control*, vol. 41, pp. 71–93, 2016.

[3] W. Song, Y. Li, and S. Tong, "Fuzzy finite-time H_∞ hybrid-triggered dynamic positioning control of nonlinear unmanned marine vehicles under cyber-attacks," *IEEE Transactions on Intelligent Vehicles*, vol. 9, no. 1, pp. 970 – 980, 2024.

[4] H. Zheng, "Coordination of Waterborne AGVs," Ph.D. dissertation, Delft University of Technology, Delft, The Netherlands, 2016.

[5] H. Zheng, W. Xu, D. Ma, and F. Qu, "Dynamic rolling horizon scheduling of waterborne AGVs for inter terminal transportation: Mathematical modeling and heuristic solution," *IEEE Transactions on Intelligent Transportation Systems*, vol. 23, no. 4, pp. 3853 – 3865, 2022.

[6] A. Bakdi, I. K. Glad, and E. Vanem, "Testbed scenario design exploiting traffic big data for autonomous ship trials under multiple conflicts with collision/grounding risks and spatio-temporal dependencies," *IEEE Transactions on Intelligent Transportation Systems*, vol. 22, no. 12, pp. 7914–7930, 2021.

[7] IMO, "Regulatory scoping exercise for the use of maritime autonomous surface ships (MASS)," International Maritime Organization, London, UK, Tech. Rep. MSC 100/WP.8, 2018.

[8] R. R. Negenborn, F. Goerlandt, T. A. Johansen, P. Slaets, O. A. Valdez Banda, T. Vanelslander, and N. P. Ventikos, "Autonomous ships are on the horizon: Here's what we need to know," *Nature*, vol. 615, no. 7950, pp. 30–33, 2023.

[9] Maritime Robotics, "World's first uncrewed freight route at sea in the trondheimsfjord," `https://www.maritimerobotics.com/post/world-s-first-uncrewed-freight-route-at-sea-in-the-trondheimsfjord`, 2023, [Online; accessed '1-August-2023].

[10] China Industrial Press, "56 ASVs in swarms into south China sea," `https://military.china.com/news/13004177/20230724/45210776.html`, 2023, [Online; accessed '1-August-2023].

[11] S. T. Pribyl, "Autonomous vessels in US shipping: Following the northern European lead," *TR News*, no. 334, pp. 14–19, 2021.

[12] A. Colling, E. van Hassel, R. Hekkenberg, M. Flikkema, W. Nzengu, B. Ramne, E. van der Linden, and N. Kreukniet, "NOVIMAR: Deliverable 1.7: Vessel train handbook," University of Antwerpe, Antwerpe, Belgium, Tech. Rep., 2021.

[13] V. Bolbot, G. Theotokatos, E. Boulougouris, L. Wennersberg, H. Nordahl, Ø. J. Rødseth, J. Faivre, and M. M. Colella, "Paving the way toward autonomous shipping development for european waters–the AUTOSHIP project," in *Autonomous Ships.* Royal Institution of Naval Architects, 2020.

[14] European Union's Horizon 2020 Research and Innovation Program, "MOSES," `https://moses-h2020.eu/`, 2020, [Online; accessed '8-September-2023].

[15] ——, "AEGIS – advanced, efficient and green intermodal systems," `https://aegis.autonomous-ship.org/`, 2020, [Online; accessed '8-September-2023].

[16] G. Avicola, "No manning required ship (NOMARS)," `https://www.darpa.mil/program/no-manning-required-ship`, 2020, [Online; accessed '8-September-2023].

[17] C. Kent, "Sea train," `https://www.darpa.mil/program/sea-train`, 2020, [Online; accessed '8-September-2023].

[18] T. W. Harker, M. M. Gilday, and D. H. Berger, "Department of navy unmanned campaign framework," Department of Navy, Washigton, USA, Tech. Rep., 2021.

[19] Z. H. Munim, M. Imset, O. Faury, M. Sukke, and H. Kim, "Public perception on safety of autonomous ferry in the Norwegian context," in *Proceedings of the IEEE International Conference on Industrial Engineering and Engineering Management*, Kuala Lumpur, Malaysia, 2022, pp. 1027–1032.

[20] The Dutch maritime sector, "SMASH! – Netherlands forum for smart shipping," `https://smashnederland.nl/en/about-smash/`, 2023, [Online; accessed '8-September-2023].

[21] The Danish maritime community, "Shipping lab – Driving the future maritime technology," `https://shippinglab.dk/`, 2019, [Online; accessed '8-September-2023].

[22] One Sea Association, "One Sea," `https://www.one-sea.org/`, 2023, [Online; accessed '15-July-2024].

[23] T. SUZUKI, "Challenge of technology development through meguri 2040," *ClassNK technical journal*, vol. 3, no. 1, pp. 51–58, 2021.

[24] T. T. D. Nguyen, S. Lee, H. K. Yoon *et al.*, "An experimental study on hydrodynamic forces of Korea autonomous surface ship in various loading condition," *Journal of Navigation and Port Research*, vol. 46, no. 2, pp. 73–81, 2022.

[25] C. Grayling and N. Ghani, "Maritime 2050: Navigating the future," The Department of Transport of UK, London, UK, Tech. Rep., 2019.

[26] R. O'Dwyer, "Maritime Innovation Lab opens in Singapore," https://smartmaritimenetwork.com/2019/04/09/maritime-innovation-lab-opens-in-singapore/, 2019, [Online; accessed '8-September-2023].

[27] S. Campbell, W. Naeem, and G. W. Irwin, "A review on improving the autonomy of unmanned surface vehicles through intelligent collision avoidance manoeuvres," *Annual Reviews in Control*, vol. 36, no. 2, pp. 267–283, 2012.

[28] D. Nađ, N. Mišković, and F. Mandić, "Navigation, guidance and control of an overactuated marine surface vehicle," *Annual Reviews in Control*, vol. 40, pp. 172–181, 2015.

[29] F. Goerlandt, "Maritime autonomous surface ships from a risk governance perspective: Interpretation and implications," *Safety science*, vol. 128, p. 104758, 2020.

[30] Z. Peng, J. Wang, D. Wang, and Q.-L. Han, "An overview of recent advances in coordinated control of multiple autonomous surface vehicles," *IEEE Transactions on Industrial Informatics*, vol. 17, no. 2, pp. 732–745, 2020.

[31] C. Barrera, I. Padron, F. Luis, and O. Llinas, "Trends and challenges in unmanned surface vehicles (USV): From survey to shipping," *TransNav: International Journal on Marine Navigation and Safety of Sea Transportation*, vol. 15, no. 1, pp. 135–142, 2021.

[32] J. Liu, X. Yan, C. Liu, A. Fan, and F. Ma, "Developments and applications of green and intelligent inland vessels in China," *Journal of Marine Science and Engineering*, vol. 11, no. 2, p. 318, 2023.

[33] Y. Qiao, J. Yin, W. Wang, F. Duarte, J. Yang, and C. Ratti, "Survey of deep learning for autonomous surface vehicles in marine environments," *IEEE Transactions on Intelligent Transportation Systems*, vol. 24, no. 4, pp. 3678–3701, 2023.

[34] M. Li, J. Zhang, Y. Ma, F. Ma, J. Liu, C. Liu, and X. Yan, "The practice and outlook of intelligent navigation of ships in China," in *Advances in Maritime Technology and Engineering*. CRC Press, 2024, pp. 9–17.

[35] H. Zheng, R. R. Negenborn, and G. Lodewijks, "Survey of approaches for improving the intelligence of marine surface vehicles," in *Proceedings of the 16th International Conference on Intelligent Transportation Systems*, Den Haag, The Netherlands, 2013, pp. 1217–1213.

[36] R. Skjetne, "The maneuvering problem," Ph.D. dissertation, Norwegian University of Science and Technology, Trondheim, Norway, 2005.

[37] A. L. S. Bjørn Johan Vartdal, Rolf Skjong, "Remote-controlled and autonomous ships," DNV GL – Maritime, Hamburg, Germany, Tech. Rep., 2018.

[38] M. Schiaretti, L. Chen, and R. R. Negenborn, "Survey on autonomous surface vessels: Part i-a new detailed definition of autonomy levels," in *Proceedings of the 8th International Conference on Computational Logistics*, Southampton, UK, 2017, pp. 219–233.

[39] SAE, "Taxonomy and definitions for terms related to driving automation systems for on-road motor vehicles," Society of Automotive Engineers, Warrendale, USA, Tech. Rep. J3016_202104, 2014.

[40] C. Liu, X. Chu, W. Wu, S. Li, Z. He, M. Zheng, H. Zhou, and Z. Li, "Human–machine cooperation research for navigation of maritime autonomous surface ships: A review and consideration," *Ocean Engineering*, vol. 246, p. 110555, 2022.

[41] G. Longo, A. Orlich, A. Merlo, and E. Russo, "Enabling real-time remote monitoring of ships by lossless protocol transformations," *IEEE Transactions on Intelligent Transportation Systems*, vol. 24, no. 7, pp. 7285–7295, 2023.

[42] K. Wróbel, "Searching for the origins of the myth: 80% human error impact on maritime safety," *Reliability Engineering & System Safety*, vol. 216, p. 107942, 2021.

[43] R. Kari and M. Steinert, "Human factor issues in remote ship operations: Lesson learned by studying different domains," *Journal of Marine Science and Engineering*, vol. 9, no. 4, p. 385, 2021.

[44] K. Wróbel, M. Gil, and J. Montewka, "Identifying research directions of a remotely-controlled merchant ship by revisiting her system-theoretic safety control structure," *Safety Science*, vol. 129, p. 104797, 2020.

[45] J. Curcio, J. Leonard, and A. Patrikalakis, "SCOUT - a low cost autonomous surface platform for research in cooperative autonomy," in *Proceedings of OCEANS 2005 Marine Technology Systems*, Washington, DC, USA, 2005, pp. 725–729.

[46] P. M. Newman, "Moos-mission orientated operating suite," *Massachusetts Institute of Technology, Tech. Rep*, vol. 2299, no. 08, 2008.

[47] G. Bruzzone, G. Bruzzone, M. Bibuli, and M. Caccia, "Autonomous mine hunting mission for the charlie USV," in *Proceedings of 2011 OCEANS*, Santander, Spain, 2011, pp. 1–6.

[48] P. Gomes, C. Silvestre, A. Pascoal, and R. Cunha, "A path-following controller for the DELFIMx autonomous surface craft," in *Proceedings of 7th IFAC Conference on Manoeuvring and Control of Marine Craft*, Lisbon, Portugal, 2006, pp. 1–6.

[49] R. Skjetne, T. I. Fossen, and P. V. Kokotović, "Adaptive maneuvering with experiments for a model ship in a marine control laboratory," *Automatica*, vol. 41, no. 2, pp. 289–298, 2005.

[50] T. Perez, O. Smogeli, T. I. Fossen, and A. Sorensen, "An overview of the marine systems simulator (MSS): a simulink toolbox for marine control systems," *Modeling, Identification and Control*, vol. 27, no. 4, pp. 259–275, 2006.

[51] F. Raimondi, M. Trapanese, V. Franzitta, A. Viola, and A. Colucci, "A innovative semi-immergible USV (SI-USV) drone for marine and lakes operations with instrumental telemetry and acoustic data acquisition capability," in *In Proceedings of OCEANS 2015*, Genova, Switzerland, 2015, pp. 1–10.

[52] W. Liu, Y. Liu, and R. Bucknall, "A robust localization method for unmanned surface vehicle (USV) navigation using fuzzy adaptive kalman filtering," *IEEE Access*, vol. 7, pp. 46 071–46 083, 2019.

[53] N. Tarcai, C. Virágh, D. Ábel, M. Nagy, P. L. Várkonyi, G. Vásárhelyi, and T. Vicsek, "Patterns, transitions and the role of leaders in the collective dynamics of a simple robotic flock," *Journal of Statistical Mechanics: Theory and Experiment*, vol. 2011, no. 04, pp. 1–26, 2011.

[54] A. Haseltalab, "Control for autonomous all-electric ships: Integrating maneuvering, energy management, and power generation control," Ph.D. dissertation, Delft, the Netherlands, 2019.

[55] M. Schiaretti, L. Chen, and R. R. Negenborn, "Survey on autonomous surface vessels: Part II-Categorization of 60 prototypes and future applications," in *Proceedings of the 8th International Conference on Computational Logistics*, Southampton, UK, 2017, pp. 234–252.

[56] T. R. Torben, Ø. Smogeli, J. A. Glomsrud, I. B. Utne, and A. J. Sørensen, "Towards contract-based verification for autonomous vessels," *Ocean Engineering*, vol. 270, p. 113685, 2023.

[57] T. K. Grose, "Skip the skipper," *ASEE Prism*, vol. 31, no. 5, pp. 8–8, 2022.

[58] A. B. Martinsen, A. M. Lekkas, S. Gros, J. A. Glomsrud, and T. A. Pedersen, "Reinforcement learning-based tracking control of USVs in varying operational conditions," *Frontiers in Robotics and AI*, vol. 7, p. 32, 2020.

[59] J. Powell, "Future of shipping," in *In Proceedings of International Conference on Marine Engineering and Technology*, Muscat, Oman, 2019, pp. 175–187.

[60] W. Wang, T. Shan, P. Leoni, D. Fernández-Gutiérrez, D. Meyers, C. Ratti, and D. Rus, "Roboat II: A novel autonomous surface vessel for urban environments," in *Proceedings of IEEE/RSJ International Conference on Intelligent Robots and Systems (IROS)*, Las Vegas, USA, 2020, pp. 1740–1747.

[61] C. Gentemann, J. P. Scott, P. L. Mazzini, C. Pianca, S. Akella, P. J. Minnett, P. Cornillon, B. Fox-Kemper, I. Cetinić, T. M. Chin *et al.*, "Saildrone: Adaptively sampling the marine environment," *Bulletin of the American Meteorological Society*, vol. 101, no. 6, pp. E744–E762, 2020.

[62] K. J. Casola, "System architecture and operational analysis of medium displacement unmanned surface vehicle sea hunter as a surface warfare component of distributed lethality," Ph.D. dissertation, Naval Postgraduate School, Monterey, USA, 2017.

[63] B.-B. Hu, H.-T. Zhang, B. Liu, H. Meng, and G. Chen, "Distributed surrounding control of multiple unmanned surface vessels with varying interconnection topologies," *IEEE Transactions on Control Systems Technology*, vol. 30, no. 1, pp. 400–407, 2021.

[64] Y. Peng, Y. Yang, J. Cui, X. Li, H. Pu, J. Gu, S. Xie, and J. Luo, "Development of the USV 'JingHai-I' and sea trials in the southern yellow sea," *Ocean engineering*, vol. 131, pp. 186–196, 2017.

[65] X. Lin, Y. Wang, and M. Cui, "Comparison of three data-driven identification methods and experimental testing on a Yunzhou unmanned surface vehicle," in *Proceedings of the 9th Data Driven Control and Learning Systems Conference*, Liuzhou, China, 2020, pp. 1131–1135.

[66] A. Haseltalab, V. Garofano, M. AFZAL, N. Faggioni, S. Li, J. Liu, F. Ma, M. Martelli, Y. Singh, P. Slaets *et al.*, "The collaborative autonomous shipping experiment (CASE): Motivations, theory, infrastructure, and experimental challenges," in *Proceedings of the International Ship Control Systems Symposium*, Delft, Netherlands, 2020, pp. 1–16.

[67] Y. Lu, H. Niu, A. Savvaris, and A. Tsourdos, "Verifying collision avoidance behaviours for unmanned surface vehicles using probabilistic model checking," *IFAC-Papers OnLine*, vol. 49, no. 23, pp. 127–132, 2016.

[68] N. Miskovic, D. Nad, and I. Rendulic, "Tracking divers: An autonomous marine surface vehicle to increase diver safety," *IEEE Robotics & Automation Magazine*, vol. 22, no. 3, pp. 72–84, 2015.

[69] A. Dallolio, H. B. Bjerck, H. A. Urke, and J. A. Alfredsen, "A persistent sea-going platform for robotic fish telemetry using a wave-propelled USV: Technical solution and proof-of-concept," *Frontiers in Marine Science*, vol. 9, p. 857623, 2022.

[70] K. Zwolak, B. Simpson, B. Anderson, E. Bazhenova, R. Falconer, T. Kearns, H. Minami, J. Roperez, A. Rosedee, H. Sade, N. Tinmouth, R. Wigley, and Y. Zarayskaya, "An unmanned seafloor mapping system: The concept of an AUV integrated with the newly designed USV SEA-KIT," in *In Proceedings of OCEANS*, Aberdeen, UK, 2017, pp. 1–6.

[71] J.-H. Choi, J.-Y. Jang, and J. Woo, "A review of autonomous tugboat operations for efficient and safe ship berthing," *Journal of Marine Science and Engineering*, vol. 11, no. 6, p. 1155, 2023.

I

Advanced Navigation, Guidance, and Control Methods Towards Autonomy

CHAPTER 2

USV Navigation Methods and A Real-Time Object Recognition System with 3D LiDAR

2.1 INTRODUCTION

NAVIGATION refers to the decision-making process of guiding the safe operation of vessels, including adherence to rules, handling emergency situations, and maneuvering in complex environments [1]. From the navigation system, the USV position, speed, water depth, environmental data, and other marine-specific measurements should be obtained. For manned ships, the navigation task is mainly carried out by the Officer of the Watch (OOW). However, for USVs, this task is achieved by the onboard navigation sensors as well as the environmental perception, modeling, and state estimation algorithms.

In this chapter, we present a navigation system framework for USVs and explores various USV navigation perception methods. Specifically, it investigates maritime object recognition based on 3D LiDAR and analyzes the architecture of a typical USV perception system equipped with 3D LiDAR. Furthermore, the chapter discusses the hardware and software implementations of real-time object recognition using 3D point clouds, including sensor selection, computing platforms, and algorithm optimization strategies. Finally, experimental verification is conducted to evaluate the effectiveness of the proposed methods in real-world maritime environments, providing reliable perception capabilities for USV autonomous navigation and obstacle avoidance.

2.2 USV NAVIGATION SYSTEM FRAMEWORK

A USV navigation system is composed of different sensors and devices for the functions of data acquisition, communication, interaction, etc. Usually, these data will be used together with an Electronic Navigation Chart (ENC). The USV navigation

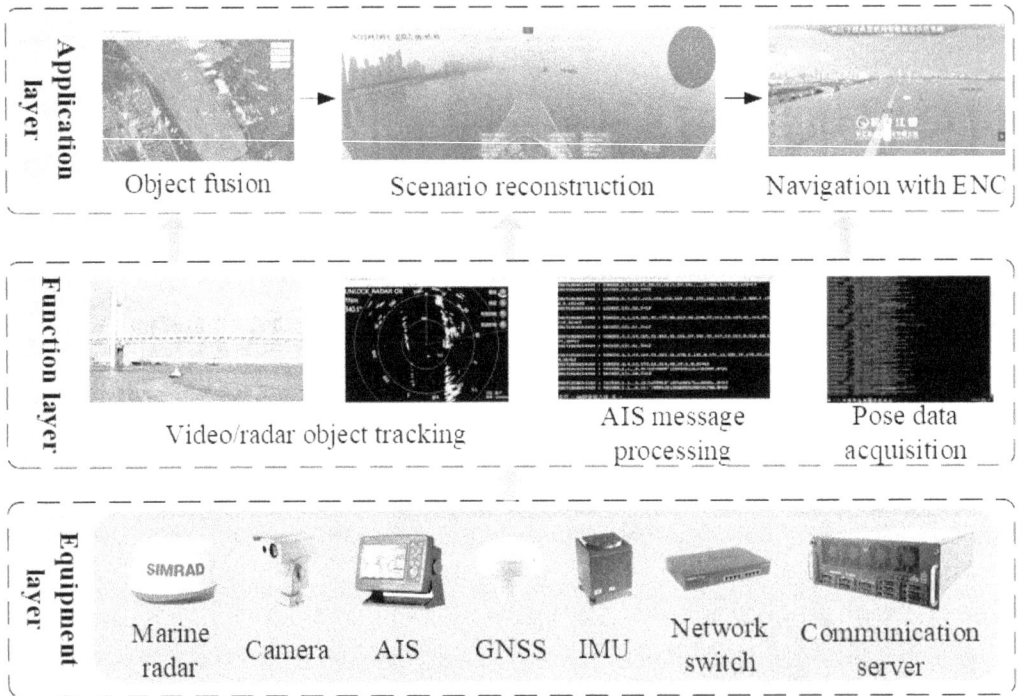

Figure 2.1: The three-layer USV hierarchical navigation system.

system can be generally divided into to three layers, i.e., the equipment layer, function layer and application layer, as shown in Fig. 2.1.

- On the equipment layer, there are the own ship perception devices (e.g., Global Navigation Satellite System (GNSS), Inertial Measurement Unit (IMU) [2], compass, etc.), environment perception devices (e.g., radar, Automatic Identification System (AIS), LiDAR [3], camera [4], millimeter-wave radar, meteorological and hydrological instruments, etc.), and the communication/computing server for the purposes of data collection, exchange and processing.

- On the function layer, the data and information from the equipment layer are analyzed with relevant technologies and methods, as discussed in the next subsection, to obtain the accurate own ship states and environmental information (e.g., other ships, obstacles, flow velocity, etc.).

- On the application layer, the USV navigation scenario is reconstituted utilizing the data from the function layer, which subsequently allows for the generation of the ENC.

The autonomous navigation with the complete perception and situation awareness as well as the following planning and control can all be carried out based on the ENC [5].

The main task of the navigation system is to integrate the information from multiple sensors to estimate the USV states and perceive the surrounding environment. According to the available navigation sensors, state estimation methods can be categorized roughly into the traditional GNSS-IMU-based approaches [2] and the alternative reactive sensor-based approaches [6]. The traditional GNSS-IMU-based state estimation methods are relatively more mature. Most commercial GNSS and IMU systems can offer estimates of the position, orientation, and velocity with high resolution and accuracy. However, the marine environment suffers from the deteriorated performance and high cost using GNSS. In GNSS-restricted scenarios, one way is to rely on the dead reckoning [1], which is, however, not reliable for long distance navigation. The other way is to use reactive sensors such as LiDAR, radar, sonar, and cameras to estimate the USV states, indirectly [7, 8].

2.3   USV NAVIGATION PERCEPTION METHODS

For safe navigation, the perception of the environment is also important. Due to the low cost, cameras have been widely used for USVs. Cameras can capture the outlines, textures, and color distributions of the targets, and thus have advantages in classifying targets [9, 4]. The deep learning technique is employed for determining the heading of vessels in images which are then tracked with the Kalman filter in [9]. More advanced stereo cameras can further provide the depth information. Pairs of stereo cameras are used in [4] for wide-angle perception of the USV environments. However, achieving high-precision three-dimensional information of targets is still challenging. Moreover, cameras are sensitive to the environment conditions, e.g., the illumination, rain, snow, and fog, which are typical in the marine environment. Therefore, radars with point cloud data are also widely used, including the millimeter-wave radar and LiDAR. Millimeter-wave radar has a longer wavelength, allowing it to penetrate harsh weather conditions. Compared to the millimeter-wave radar, LiDAR can provide high-resolution of data at the sacrifice of perception range. For the short range intelligent vessel berthing application [3, 10], LiDAR has shown outstanding performance. The high-precision point cloud maps of coastlines are created and the Simultaneous Localization and Mapping (SLAM) can be achieved during the autonomous berthing task.

Mostly, for complement and redundancy, multi-sensor data are combined and fused for the navigation of USVs [11, 12]. Existing image-point cloud fusion methods are predominantly designed as unidirectional fusion neural networks tailored for specific tasks. Variants like LidCamNet [13] and progressive LiDAR adaptation-aided road detection [14] project the point clouds onto the image plane for the unidirectional fusion with images. Most of these methods uniformly represent images and point clouds by discarding the data original structure, resulting in a lack of generality in multiple tasks. The bidirectional fusion network is proposed in [15] preserving the original structures of both images and point clouds.

In maritime scenarios, due to the vast navigation areas at sea, the targets to be identified and tracked are often sparse. For the simultaneous tracking of sparse targets by multiple sensors, disparities in the target's position exist in both time and space [16]. Additionally, different sensors exhibit variations in data formats and processing methods. In such circumstances, commonly employed algorithms include the k-Nearest Neighbor (K-NN) algorithm [17], federated Kalman filtering [18], Joint Probability Data Association (JPDA) [19] are applicable. For high-traffic maritime areas, such tracking of targets introduces challenges related to data overlap and trajectory intersection. Solutions to these issues include the improved JPDA algorithms, multiple hypothesis tracker [20], and the random finite set [21] theory, among others. The artificial intelligence techniques have also shown tremendous popularity in the USV situation awareness problems [22] in recent years.

2.4 MARITIME OBJECT RECOGNITION WITH 3D LIDAR

The perception devices commonly used in current USVs exhibit deficiencies in recognizing surrounding obstacles during navigation. Maritime radar, for instance, suffers from short-range blind spots and inaccurate distance measurements. Visual sensors encounter issues with all-weather operation and direct distance measurement capabilities. Millimeter-wave radar presents problems of low perception resolution and measurement blind zones. These issues collectively present significant difficulties in real-time target detection and obstacle avoidance for USVs. Three-dimensional Light Detection and Ranging (LiDAR) technology provides high-resolution, all-weather, and interference-resistant real-time 3D point cloud data of the surrounding environment. Processing the generated 3D point cloud allows for real-time obstacle target status assessment within the scene. This technology has found extensive application in fields such as unmanned aerial vehicles, autonomous vehicles, and robot obstacle avoidance. However, the detection range of 3D LiDAR is limited (less than 300 m), which poses limitations for obstacle perception in medium to large-sized vessels. As small-scale waterborne vehicles, USVs exhibit excellent maneuverability and are well-suited for the environmental perception needs of 3D LiDAR. Consequently, 3D LiDAR becomes a crucial means for identifying surface obstacles and targets, playing a key role in achieving autonomous navigation for USVs.

Currently, LiDAR has found applications in environmental perception for USVs. In 2016, during the International RoboBoat Competition held by the Association for Unmanned Vehicle Systems International (AUVSI), participating vessels commonly employed LiDAR as target detection sensors [23]. In 2018, the Maersk Group in Denmark installed LiDAR and other perception devices on their container ships to aid in navigational environmental perception and motion feature recognition, providing data support and situational awareness foundation for collision avoidance research. Additionally, 3D LiDAR has been used for berth and dock recognition during the mooring and unmooring process of ships. Evidently, LiDAR's applications in intelligent perception systems for maritime vessels are becoming increasingly extensive.

In comparison to vehicles, aircraft, and robots, the application of LiDAR on USVs is still relatively immature [29]. Previous works have conducted research on LiDAR-based target recognition for USVs, proposing target recognition and classification algorithms that consider vessel motion and waterborne target characteristics. However, these algorithms have not been tested in actual operational environments for USVs.

[26] introduced a LiDAR-based obstacle avoidance method for USV path following in harbor environments, but the discussion on LiDAR point cloud processing is limited. Addressing the aforementioned limitations, this paper designs a real-time target recognition system for USVs based on 3D LiDAR. The paper outlines the design of system hardware, software, and communication structure in accordance with system requirements, investigates key technologies including point cloud processing, target segmentation, and remote interaction of point cloud images, constructs an experimental platform for real-time 3D LiDAR target recognition on USVs, and conducts testing and verification of system algorithms and performance.

2.5   TYPICAL USV PERCEPTION SYSTEM WITH 3D LIDAR

During the navigation process, environmental obstacles pose significant safety risks to USVs, and having real-time detection capability for environmental obstacles is fundamental for achieving autonomous navigation of USVs. When operating autonomously, USVs need to acquire precise information about the position, azimuth angle, size, and other status details of obstacles within a close range (¡300 m) in order to perform obstacle avoidance maneuvers. Therefore, the designed USV-based 3D LiDAR target recognition system will effectively address the issue of recognizing obstacles at short distances.

The obstacle recognition process of the designed USV-based 3D LiDAR target recognition system (referred to as the USV recognition system) is as follows:

1) Obtaining the USV's position, velocity, attitude angles, and other information through the Global Navigation Satellite System (GNSS), electronic compass, and attitude sensors. The calibrated coordinates of the LiDAR point cloud are then calculated.

2) After processing the 3D LiDAR point cloud through filtering and rasterization techniques, real-time segmentation and target recognition of the rasterized points are achieved using an eight-neighbor algorithm.

3) Utilizing TCP/IP transmission, the results of obstacle recognition are sent to the navigation decision unit and the remote control unit.

The USV-based 3D LiDAR target recognition system primarily consists of the USV itself, perception devices, an onboard industrial PC, and a shore-based client. The system architecture is illustrated in Fig. 2.2.

The functions of each module in the system are as follows:

1) 3D LiDAR, Cameras, and Image Perception Devices: These devices are responsible for detecting the status information of obstacles within the navigation area of the USV and sending this information to the onboard industrial PC.

2) GNSS, Electronic Compass, and Attitude Sensors: These state perception devices are utilized to detect the USV's attitude angles, acceleration, and real-time geographic coordinates during navigation. The information collected from these devices is continuously transmitted back to the onboard industrial PC, facilitating accurate coordinate positioning of the USV and providing data for attitude correction of the LiDAR measurements.

Sensing device

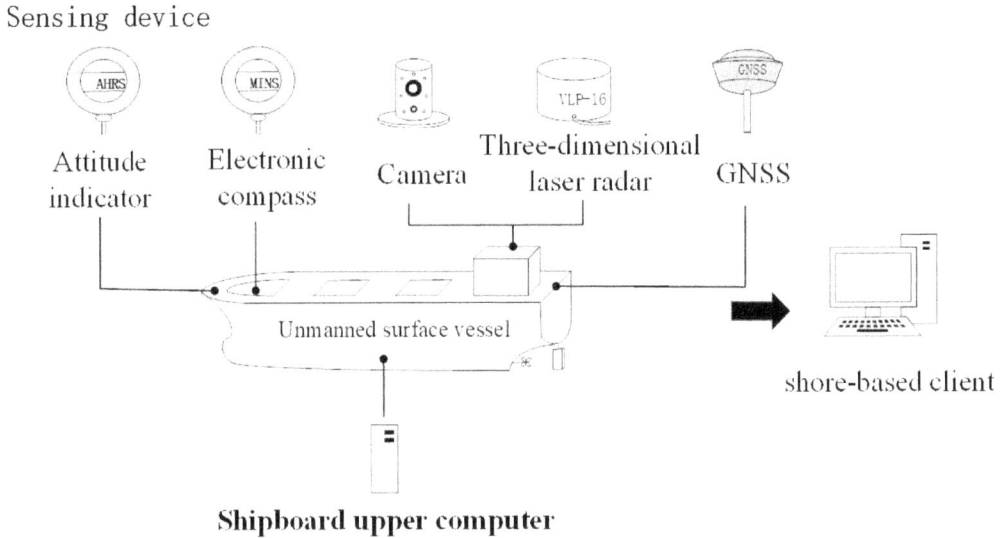

Figure 2.2: Structure composition of the platform.

3) Onboard Industrial PC: The onboard industrial PC receives data from the perception devices and the 3D LiDAR mounted on the USV. It processes this data in real-time to generate obstacle recognition results concerning the obstacles around the USV.

4) Shore-Based Client: The shore-based client serves as an interface for displaying real-time information about the USV's status during navigation, as well as obstacle recognition results. Additionally, it enables remote control of the USV by shore-based operators.

2.6 IMPLEMENTATIONS: HARDWARE AND SOFTWARE

2.6.1 Hardware implementations

The hardware structure of the USV recognition system primarily includes the hull, power propulsion module, 3D LiDAR, industrial PC, attitude sensors, GNSS, electronic compass, power supply module, communication module, and remote control module, among others. The main components of the USV recognition system are described below, as presented in Table 2.1.

The data transmission structure of the unmanned boat recognition system is shown in Fig. 2.3. Firstly, the remote control unit sends navigation instructions to the unmanned boat and sets basic parameters such as waypoints. Secondly, the onboard computer calculates control commands based on the received instructions and environmental obstacle information, and sends them to the motion drive unit to activate the propeller and rudder. Thirdly, the 3D LiDAR detects the surrounding information of the unmanned boat, while the GNSS receiver, electronic compass, and attitude sensor collect navigation information such as latitude, longitude, and attitude data, and transmit the 3D LiDAR point cloud data and navigation status data

TABLE 2.1 Main components of the platform

Device Name	Main Function
USV hull	Surface vehicle
Power propulsion module	Controls the propellers and rudders
3D LiDAR	Obtaining three-dimensional point cloud data of the surrounding environment of the USV
Shipborne upper computer	Utilized for the fusion, processing, and interaction of LiDAR point cloud data and other data, as well as sending control commands
Attitude indicator	Acquiring real-time attitude angles, acceleration, and other data of the USV
GNSS	Acquiring Real-Time Latitude, Longitude, and Velocity Data of the USV
Electron industry	Obtaining accurate heading data of the USV
Power supply module	Providing power to all equipment of the USV
Communication module	Ensuring proper internal communication within the USV and functional communication between the vessel and shore-based systems
Remote control module	Real-time remote monitoring of the operational status of the USV

to the onboard computer in real time. Finally, the onboard computer transmits the perception device information to the remote control unit for data display and storage.

2.6.2 Software implementations

(1) Software architecture and functions

On the Microsoft Windows 10 operating system, utilizing platforms such as Microsoft Visual Studio 2017, Point Cloud Library (PCL) 1.9.1, and Qt 5.9.2, a software application titled "VLP16Process 1.0" (referred to as VLP 1.0) has been developed. This software is designed for the purpose of three-dimensional LiDAR target recognition on USVs, specifically targeting the Velodyne company's 16-line LiDAR sensor (VLP16). VLP 1.0 provides real-time functionalities for reading, processing, recognizing, displaying, storing, and outputting three-dimensional LiDAR point cloud data.

The functional framework of the VLP 1.0 software is illustrated in Fig. 2.4.

1) Point Cloud Display Area: Located at the top of the system window, this area displays the three-dimensional point cloud image (on the left) of obstacles around the experimental vessel collected by the ship-mounted 3D LiDAR. It also displays the rasterized image (on the right) after identifying the contours of these obstacles.

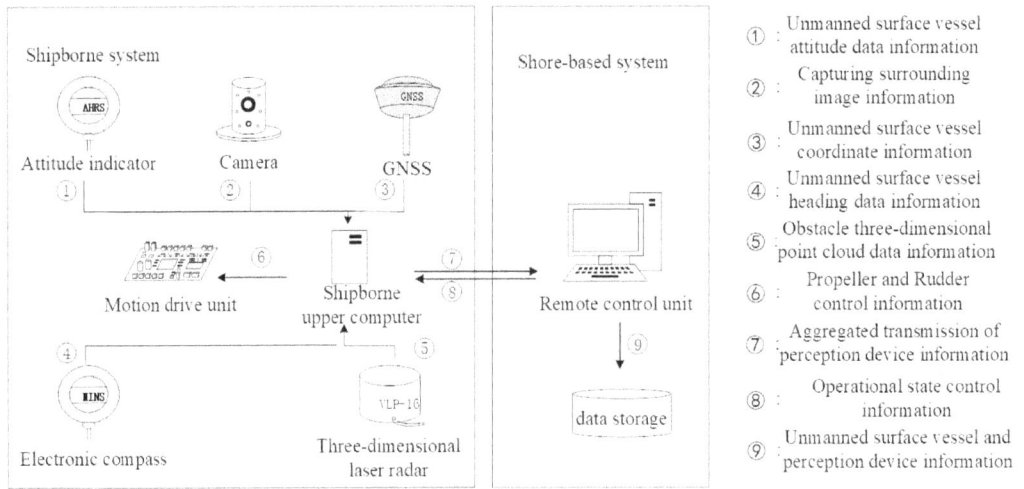

Figure 2.3: Data transmission structure.

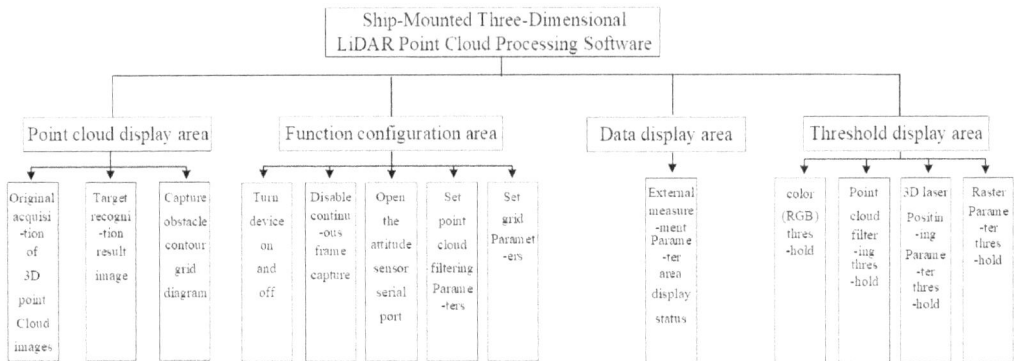

Figure 2.4: VLP1.0 software functional framework.

2) Function Configuration Area: Each button in this area corresponds to a specific function, as detailed in Table 2.2.

TABLE 2.2   Software button function

Device Name	Main Function
Open LiDAR	Turn on the 3D laser object recognition device
Stop Receive	Stop receiving 3D laser values
Filter Set	Set point cloud filtering parameters
Grid Set	Grid parameter Settings
Stop Display	Stop capturing consecutive frames
Open AttiCom	Open the attitude sensor serial port

3) Data Display Area: The data display area is intended for presenting real-time variations in state variables during the vessel's navigation process. This encompasses

parameters such as the vessel's navigation speed ("Speed"), heading or yaw angle ("Heading"), pitch angle ("Pitch"), and roll angle ("Roll").

4) Status Output Area: Positioned at the bottom-left corner of the window, the status output area showcases various states, including radar connectivity, data storage, and communication.

(2) Software development

The VLP 1.0 software interface, as depicted in Fig. 2.5, comprises several functional areas, including original three-dimensional point cloud display, display of target recognition results (distinct color-coded grids representing various obstacle targets), software operational status display, system and parameter settings, and vessel status display. This software facilitates functionalities such as point cloud preprocessing, point cloud segmentation, obstacle grid marking, point cloud visualization, and configuration. It achieves a processing speed exceeding 5 Hz and is capable of transmitting obstacle recognition results to the navigation decision computation unit via the TCP/IP protocol, enabling real-time collision avoidance for intelligent vessels.

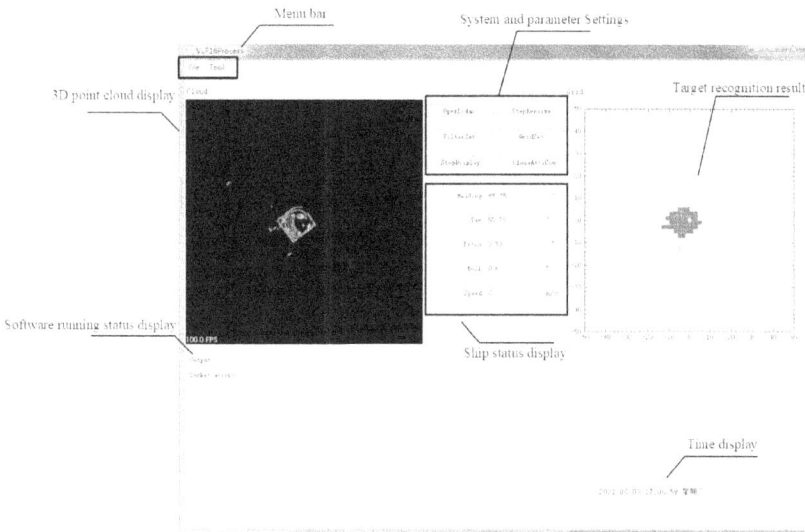

Figure 2.5: VLP 1.0 point cloud processing software interface.

2.7   REAL-TIME OBJECT RECOGNITION BY 3D POINT CLOUD

2.7.1   Point cloud data processing

The point cloud data preprocessing process involves coordinate transformation and calibration, point cloud filtering, and grid representation, among other steps.

1) Coordinate Transformation and Calibration

During point cloud data processing for the USV, four coordinate systems need to be considered for coordinate transformation [28]: The LiDAR coordinate system $l=(\rho,\alpha,\omega)$, which is a spherical coordinate system with the center of the LiDAR as the origin. The body-fixed coordinate system $b=(x,y,z)$, which is a Cartesian coordinate system with the center of the LiDAR as the origin, the longitudinal section of the vessel as the x-axis, and the direction perpendicular to the longitudinal section as the y-axis. The horizontal plane coordinate system $b_0 = (x_0, y_0, z_0)$, which is a Cartesian coordinate system with the center of the LiDAR as the origin, the north direction as the x-axis, and the east direction as the y-axis. The global (geodetic) coordinate system $n = (X, Y, Z)$, as shown in Fig. 2.6. The transformation involves converting the

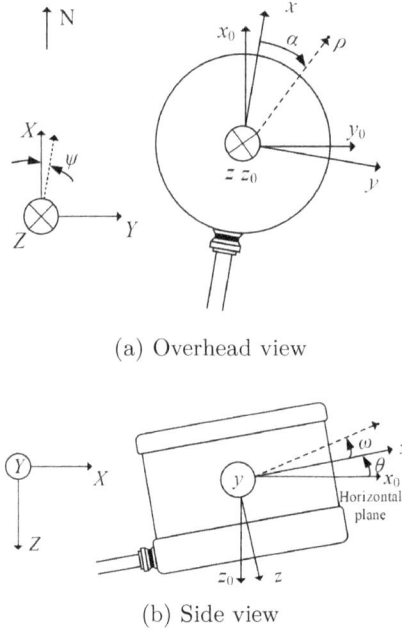

(a) Overhead view

(b) Side view

Figure 2.6: Coordination systems for USV.

raw LiDAR point cloud from its spherical coordinate form (ρ, α, ω) to the Cartesian coordinates in the body-fixed coordinate system of the vessel:

$$\begin{cases} x = \rho \cos \omega \cos \alpha \\ y = \rho \cos \omega \sin \alpha \ , \\ z = \rho \sin \omega \end{cases} \tag{2.1}$$

In the equation, ρ represents the spherical radius in spherical coordinates;α represents the azimuth angle in spherical coordinates; ω represents the elevation angle in spherical coordinates.

During the scanning process of the LiDAR, the roll, pitch, and yaw motions of the USV can cause discrepancies between the actual scanned point cloud coordinates and the coordinates of the vessel in a stable horizontal state (horizontal plane coordinate system b_0). Thus, it becomes necessary to compensate and correct the point cloud coordinates for vessel motion. Let ϕ, θ, and ψ denote the roll, pitch, and yaw angles of the vessel, respectively. The corresponding rotation matrices are R_x, ϕ, R_y, θ and

R_z, ψ:

$$R_{x,\phi} = \begin{bmatrix} 1 & 0 & 0 \\ 0 & \cos\phi & -\sin\phi \\ 0 & \sin\phi & \cos\phi \end{bmatrix},$$

$$R_{y,\theta} = \begin{bmatrix} \cos\theta & 0 & \sin\theta \\ 0 & 1 & 0 \\ -\sin\theta & 0 & \cos\theta \end{bmatrix}, \tag{2.2}$$

$$R_{z,\psi} = \begin{bmatrix} \cos\psi & -\sin\psi & 0 \\ \sin\psi & \cos\psi & 0 \\ 0 & 0 & 1 \end{bmatrix}.$$

Under the circumstance of vessel motion, the LiDAR point cloud coordinates (x, y, z) in the body-fixed coordinate system b can be corrected to coordinates (x_0, y_0, z_0) in the horizontal plane coordinate system:

$$\begin{bmatrix} x_0 \\ y_0 \\ z_0 \end{bmatrix} = R_{x,\phi} R_{y,\theta} R_{z,\psi} \begin{bmatrix} x \\ y \\ z \end{bmatrix}, \tag{2.3}$$

By adding (x_0, y_0, z_0) to the coordinates of the LiDAR's center point in the global coordinate system n, you can obtain the coordinates of each LiDAR point cloud in the n coordinate system. Assuming at time t, the coordinates of a point cloud in the l coordinate system are (ρ, α, ω), its corresponding coordinates in the n coordinate system are (X, Y, Z):

$$\begin{bmatrix} X \\ Y \\ Z \end{bmatrix} = \begin{bmatrix} X_L \\ Y_L \\ Z_L \end{bmatrix} + R_{x,\phi} R_{y,\theta} R_{z,\psi} \begin{bmatrix} \rho\cos\omega\cos\alpha \\ \rho\cos\omega\sin\alpha \\ \rho\sin\omega \end{bmatrix}, \tag{2.4}$$

In the equation, (X_L, Y_L, Z_L) represents the coordinates of the LiDAR center point in the n coordinate system at time t, which can be obtained through GNSS measurements.

2) Point cloud filtering

The filtering of the 3D LiDAR point cloud primarily involves three processes: In the first step, a pass-through filter is applied. Thresholds are set as needed for the LiDAR point cloud along the x, y, and z axes, as well as for the scanning angle. Points outside these thresholds are removed. In the second step, a voxelization algorithm is employed to down sample the point cloud data. This involves creating a 3D voxel grid for the point cloud data, and representing all points within a voxel with their centroid. This process filters out redundant data points while preserving the shape characteristics of the point cloud. Finally, an intensity threshold is set for the LiDAR point cloud. Points with intensities below this threshold are considered noise and are filtered out. These filtering processes help refine and clean the point cloud data while retaining important shape and intensity information.

3) Raster representation

Taking the Velodyne VLP-16 LiDAR as an example, which can generate up to 300,000 data points per second, processing each point individually during target detection would result in a heavy computational load and reduced real-time performance. To address this, it's necessary to convert the 3D point cloud data into a grid map for processing [24, 27]. In this chapter, a 2D grid map is established based on the LiDAR's detection range, with the LiDAR center as the origin of the grid map, as illustrated in Figs. 2.7 and 2.8. Here, R represents the LiDAR's detection range, and the grid size is related to the vessel's length. Typically, each grid contains several laser scan points. Grid attributes include grid coordinates, presence of laser return points, number of points within the grid, and average height value, among others. A grid is defined as an obstacle grid if it contains more points than a certain threshold, otherwise, it is considered a non-obstacle grid.

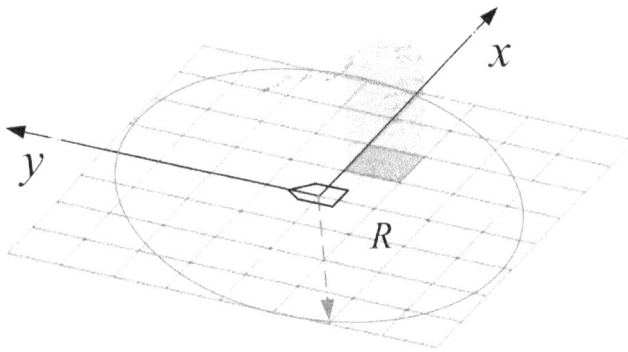

Figure 2.7: Point cloud projection

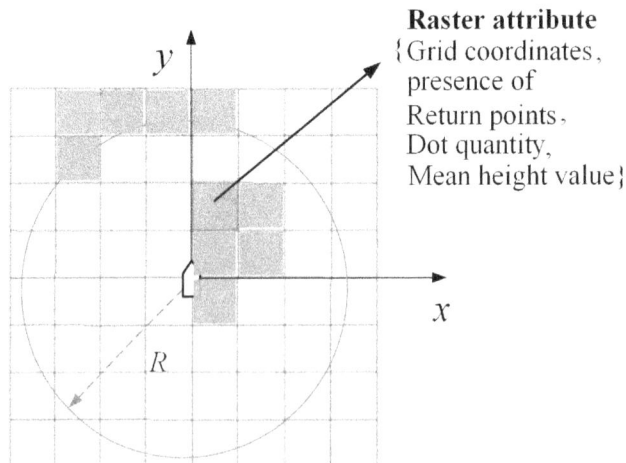

Figure 2.8: Grid map.

2.7.2 Point cloud segmentation

The eight-neighborhood search method, based on the grid structure, offers advantages such as ease of implementation and low computational requirements [25]. Hence, this book employs a fast region labeling approach using eight-neighborhood expansion for point cloud segmentation. The selected grid C serves as the current central grid, and there are eight adjacent grids in all directions, labeled as N_0, N_1, N_2, ..., N_7, as illustrated in Fig. 2.9. This method allows for efficient segmentation of the point cloud data based on the relationships between neighboring grids. The steps for im-

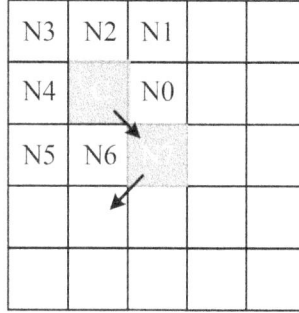

Figure 2.9: Definition of eight neighborhoods.

plementing point cloud segmentation using the eight-neighborhood approach are as follows:

1) Start from the bottom-left corner grid (with the smallest x and y coordinates) in the grid coordinate system. Traverse from left to right and from bottom to top to find the first obstacle grid, which will be the starting point for obstacle O_1.

2) Use the grid to the right of the starting point grid (N_0) as the initial direction ($C \rightarrow N_0$) and traverse its eight neighboring grids counterclockwise.

3) If any of the grids N_p (where $p = 0, 1, 2, ..., 7$) is an obstacle grid, rotate the search direction 90° clockwise (subtract 2 from the current search direction) around N_p as the center grid. This is done to avoid re-searching grids that were already searched in the previous step.

4) Repeat steps 2 and 3 to continue searching for the next obstacle grid until no other obstacle grids can be found (excluding the ones already searched). This will give you the set of obstacle grids for obstacle O_1.

5) Start from the bottom-left corner grid again to find the starting grid for the next obstacle O_2, ensuring that the starting grid for O_2 does not overlap with any of the obstacle grids for O_1. Repeat steps 2 to 4 to obtain the set of obstacle grids for obstacle O_2.

6) Continue this process to obtain the set of obstacle grids for all obstacles on the grid map. In the example, the segmented obstacle grids obtained after grid segmentation are shown in Fig. 2.10, where obstacles "1," "2," and "3" represent the segmented obstacle targets.

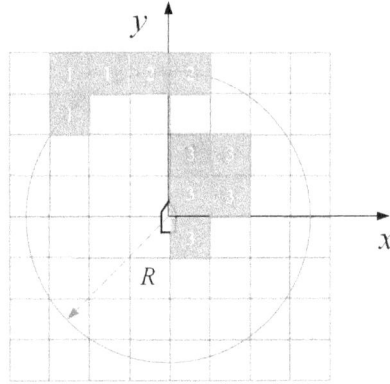

Figure 2.10: Results of barrier grid segmentation.

2.7.3 Point cloud image remote transmission technology

Considering the large real-time data volume of 3D LiDAR point clouds, it's often challenging to achieve real-time remote monitoring of the 3D point cloud data due to limitations in transmission bandwidth. To address this, the system has developed a 3D LiDAR point cloud compression technique. This technique transforms the 3D point cloud into image data and applies rotation correction to the images based on the real-time heading of the vessel. The rotation correction ensures that the top of the image corresponds to the north direction, achieving the same display effect as marine radar.

The real-time identification system for 3D LiDAR targets utilizes the shipboard computer to send 3D LiDAR data to a destination host IP via the UDP protocol. This approach enables real-time monitoring and storage of the motion process of the unmanned vessel and the surrounding obstacle information obtained by the perception devices, as shown in Fig. 2.11.

2.8 EXPERIMENTAL VERIFICATION

This book conducted autonomous navigation experiments with an unmanned vessel in an outdoor water pool environment to validate the hardware, software, and algorithm performance of the 3D LiDAR target recognition system. A 3D LiDAR and a camera were installed at the stern of the vessel to simultaneously capture the 3D point cloud data of the surrounding obstacles and video information, as depicted in Fig. 2.12. On May 1, 2021, a platform for real-time recognition of 3D LiDAR targets on unmanned vessels was established in the campus waterway of Minjiang University in Fuzhou City. The specific experimental environment is depicted in Fig. 2.13. The

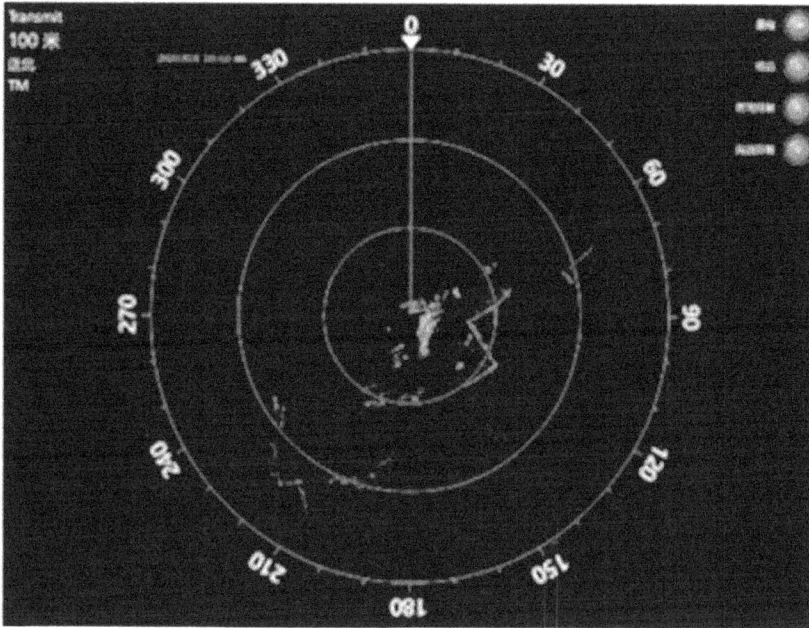

Figure 2.11: Laser point cloud image remote transmission effect.

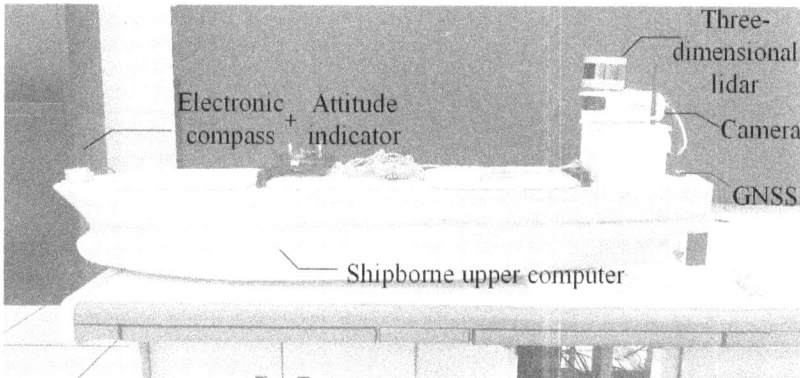

Figure 2.12: USV platform.

Velodyne VLP-16 3D LiDAR used in the experiment has an effective range of approximately 100 meters in the water environment. It operates at a scanning frequency of 5 to 20 Hz. The acquired LiDAR data is processed using a two-dimensional grid method to create a grid representation. The grid segmentation is performed using the eight-neighbor algorithm. The obtained video information and segmented grid information are presented in Figs. 2.14 and 2.15 respectively, where each color in the grid represents a specific obstacle target. By comparing the images in Figs. 2.14 and 2.15, it is evident that the system is capable of accurately identifying obstacles around the unmanned vessel and correctly categorizing the detected targets. The interface of the 3D LiDAR target recognition system reveals that the system converts the acquired 3D LiDAR data into grid image data. It employs the compass heading data from

Figure 2.13: Experimental scenario.

Figure 2.14: Video of an obstacle.

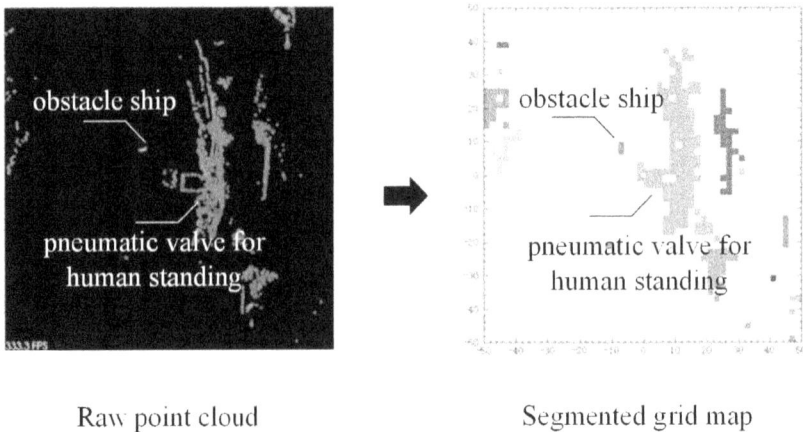

Raw point cloud   Segmented grid map

Figure 2.15: Point cloud segmentation and target recognition algorithm.

the electronic compass and the attitude angle data from the inclinometer to perform real-time rotation correction on the image, aligning the top side of the image with the north direction. Typical point cloud images and grid images at 90 s, 160 s, 240 s, and 314 s during the experiment were selected for validation, as shown in Fig. 2.16. In this figure, t represents the recorded time during the experiment, and ψ represents the vessel's heading angle at that moment. The experimental results indicate:

1) The real-time processing frequency of the LiDAR point cloud exceeds 5 Hz, meeting the requirements for unmanned vessel navigation perception and control.

2) The system effectively senses the surrounding environment of the unmanned vessel and performs angle correction on the point cloud images.

3) By matching the identified obstacle information from different moments with the captured images at those moments, it was observed that the recognized obstacle results correspond to the obstacles in the images. This validates the effectiveness of the point cloud processing and the eight-neighbor segmentation algorithm.

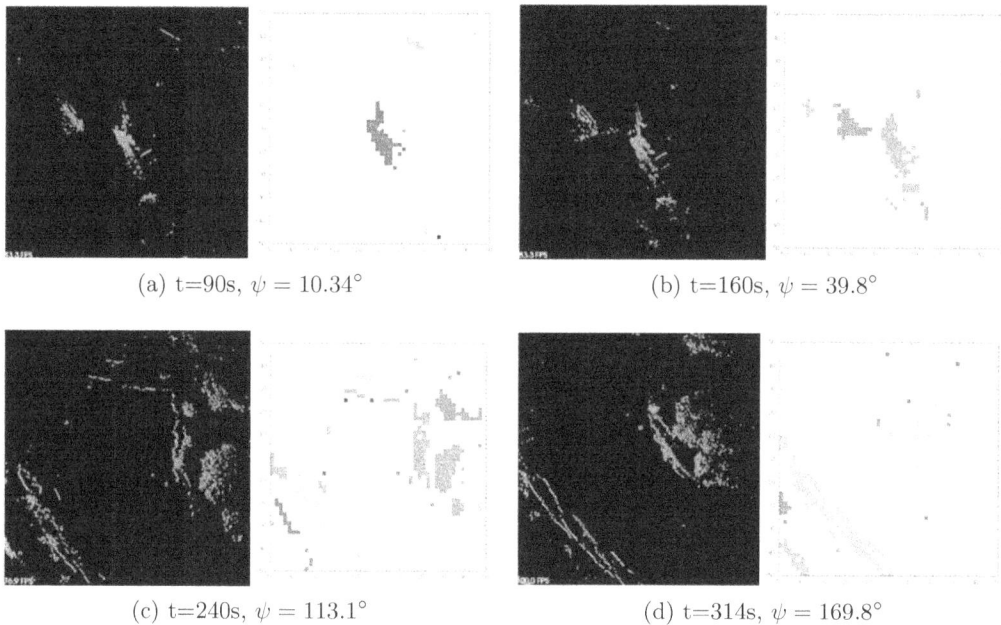

(a) t=90s, $\psi = 10.34°$

(b) t=160s, $\psi = 39.8°$

(c) t=240s, $\psi = 113.1°$

(d) t=314s, $\psi = 169.8°$

Figure 2.16: Automatic calibration and recognition of point cloud images.

2.9 CONCLUSIONS

This chapter utilizes 3D LiDAR and other perception technologies to design a real-time target recognition system for unmanned vessels in navigational environments. The system enables real-time perception and recognition of surrounding obstacles during the unmanned vessel's navigation. By constructing a real platform for the unmanned vessel's 3D LiDAR target recognition system, various functionalities and reliability of the system were thoroughly tested. The results indicate that the system reliably identifies obstacles within a range of approximately 100 meters around the unmanned vessel.

Although the effective detection range of current three-dimensional LiDAR is relatively short and cannot replace the long-distance obstacle detection capability of conventional shipborne navigation radar, the target perception capability of 3D LiDAR effectively compensates for the shortcomings of existing short-range target perception on ships. Additionally, it provides berth contour information during the autonomous berthing and unberthing processes. Furthermore, the integration of the 3D LiDAR target perception system with traditional marine radar automatic plotting enhances the vessel environmental perception capabilities.

Bibliography

[1] T. I. Fossen, *Handbook of Marine Craft Hydrodynamics and Motion Control.* West Sussex, U.K.: John Wiley and Sons Ltd., Second edition, 2021.

[2] P. Lyu, B. Wang, J. Lai, S. Bai, M. Liu, and W. Yu, "A factor graph optimization method for high-precision imu-based navigation system," *IEEE Transactions on Instrumentation and Measurement*, vol. 72, pp. 1–12, 2023.

[3] R. Sawada and K. Hirata, "Mapping and localization for autonomous ship using lidar slam on the sea," *Journal of Marine Science and Technology*, vol. 28, p. 410–421, 2023.

[4] T. Huntsberger, H. Aghazarian, A. Howard, and D. C. Trotz, "Stereo vision–based navigation for autonomous surface vessels," *Journal of Field Robotics*, vol. 28, no. 1, pp. 3–18, 2011.

[5] J. Liu, X. Yan, C. Liu, A. Fan, and F. Ma, "Developments and applications of green and intelligent inland vessels in China," *Journal of Marine Science and Engineering*, vol. 11, no. 2, p. 318, 2023.

[6] V. Vasilopoulos, G. Pavlakos, K. Schmeckpeper, K. Daniilidis, and D. E. Koditschek, "Reactive navigation in partially familiar planar environments using semantic perceptual feedback," *The International Journal of Robotics Research*, vol. 41, no. 1, pp. 85–126, 2022.

[7] H. Ma, E. Smart, A. Ahmed, and D. Brown, "Radar image-based positioning for USV under GPS denial environment," *IEEE Transactions on Intelligent Transportation Systems*, vol. 19, no. 1, pp. 72–80, 2017.

[8] J. Han, Y. Cho, and J. Kim, "Coastal SLAM with marine radar for USV operation in GPS-restricted situations," *IEEE Journal of Oceanic Engineering*, vol. 44, no. 2, pp. 300–309, 2019.

[9] W.-J. Lee, M.-I. Roh, H.-W. Lee, J. Ha, Y.-M. Cho, S.-J. Lee, and N.-S. Son, "Detection and tracking for the awareness of surroundings of a ship based on deep learning," *Journal of Computational Design and Engineering*, vol. 8, no. 5, pp. 1407–1430, 2021.

[10] H. Wang, Y. Yin, Q. Jing, Z. Cao, Z. Shao, and D. Guo, "Berthing assistance system for autonomous surface vehicles based on 3D LiDAR," *Ocean Engineering*, vol. 291, p. 116444, 2024.

[11] J. Kim, C. Lee, D. Chung, Y. Cho, J. Kim, W. Jang, and S. Park, "Field experiment of autonomous ship navigation in canal and surrounding nearshore environments," *Journal of Field Robotics*, vol. 41, no. 2, pp. 470–489, 2024.

[12] J. Han, Y. Cho, J. Kim, J. Kim, N.-s. Son, and S. Y. Kim, "Autonomous collision detection and avoidance for ARAGON USV: Development and field tests," *Journal of Field Robotics*, vol. 37, no. 6, pp. 987–1002, 2020.

[13] L. Caltagirone, M. Bellone, L. Svensson, and M. Wahde, "Lidar–camera fusion for road detection using fully convolutional neural networks," *Robotics and Autonomous Systems*, vol. 111, pp. 125–131, 2019.

[14] Z. Chen, J. Zhang, and D. Tao, "Progressive lidar adaptation for road detection," *IEEE/CAA Journal of Automatica Sinica*, vol. 6, no. 3, pp. 693–702, 2019.

[15] Q. Wang, J. Chen, J. Deng, and X. Zhang, "Pi-net: An end-to-end deep neural network for bidirectionally and directly fusing point clouds with images," *IEEE Robotics and Automation Letters*, vol. 6, no. 4, pp. 8647–8654, 2021.

[16] D. Petković, "Adaptive neuro-fuzzy fusion of sensor data," *Infrared Physics & Technology*, vol. 67, pp. 222–228, 2014.

[17] K. Ma, H. Zhang, R. Wang, and Z. Zhang, "Target tracking system for multi-sensor data fusion," in *Proceedings of the 2017 IEEE 2nd Information Technology, Networking, Electronic and Automation Control Conference (ITNEC),*, Chengdu, China, 2017, pp. 1768–1772.

[18] Z. Tian, Y. Cai, S. Huang, F. Hu, Y. Li, and M. Cen, "Vehicle tracking system for intelligent and connected vehicle based on radar and V2V fusion," in *Proceedings of the 2018 Chinese Control And Decision Conference*, Shenyang, China, 2018, pp. 6598–6603.

[19] S. He, H.-S. Shin, and A. Tsourdos, "Joint probabilistic data association filter with unknown detection probability and clutter rate," *Sensors*, vol. 18, no. 1, p. 269, 2018.

[20] S. Graebenitz, M. Mertens, and F. Belfiori, "Sequential data fusion applied to a distributed radar, acoustic and visual sensor network," in *Proceedings of the 2017 Sensor Data Fusion: Trends, Solutions, Applications (SDF),*, Bonn, Germany, 2017, pp. 1–6.

[21] H. Blom and E. Bloem, "Probabilistic data association avoiding track coalescence," *IEEE Transactions on Automatic Control*, vol. 45, no. 2, pp. 247–259, 2000.

[22] S. Thombre, Z. Zhao, H. Ramm-Schmidt, J. M. V. García, T. Malkamäki, S. Nikolskiy, T. Hammarberg, H. Nuortie, M. Z. H. Bhuiyan, S. Särkkä *et al.*, "Sensors and AI techniques for situational awareness in autonomous ships: A review," *IEEE Transactions on Intelligent Transportation Systems*, vol. 23, no. 1, pp. 64–83, 2020.

[23] L. Stanislas, M. Dunbabin, "Multimodal sensor fusion for robust obstacle detection and classification in the maritime robotx challenge," *IEEE Journal of Oceanic Engineering*, vol. 44, no. 2, pp. 343–351, 2018.

[24] S. Hrabar, "An evaluation of stereo and laser-based range sensing for rotorcraft unmanned aerial vehicle obstacle avoidance," *Journal of Field Robotics*, vol. 29, no. 2, pp. 215–239, 2012.

[25] J. Villa, J. Aaltonen, K.T. Koskinen, "Path-following with lidar-based obstacle avoidance of an unmanned surface vehicle in harbor conditions," *IEEE/ASME Transactions on Mechatronics*, vol. 25, no. 4, pp. 1812—1820, 2020.

[26] K. Miadlicki, M. Pajor, M. Sakow, "Ground plane estimation from sparse lidar data for loader crane sensor fusion system," in *Proceedings of the 22nd International Conference on Methods and Models in Automation and Robotics (MMAR)*, Międzyzdroje, Poland, 2017, pp. 717—722.

[27] C.R. Qi, W. Liu, C. Wu, H. Su, L.J. Guibas, "Frustum pointnets for 3d object detection from rgb-d data," in *Proceedings of the 2018 IEEE Conference on Computer Vision and Pattern Recognition*, Salt Lake City, USA, 2018, pp. 918—927.

[28] P. Kaur, I. Lamba, A. Gosain, "A robust method for image segmentation of noisy digital images," in *Proceedings of the 2011 IEEE International Conference on Fuzzy Systems*, 2011, pp. 1656--1663.

[29] D. Thompson, E. Coyle, J. Brown, "Efficient lidar-based object segmentation and mapping for maritime environments," *IEEE Journal of Oceanic Engineering*, vol. 44, no 2, pp. 352—362, 2019.

CHAPTER 3

USV Guidance Methods and COLREGs-Compliant Path Planning

3.1 INTRODUCTION

Autonomous guidance is one of the core technologies of USVs. Guidance is the process for guiding the path of the USV towards given waypoints [1]. It usually takes two steps for achieving this task:

(1) Planning a path in the navigation map for the USV;

(2) Guiding the USV to the planned path maneuvering along which will guide the USV to the given waypoints.

In this chapter, we first review and classify the state-of-the-art USV guidance methods. Then, a frequently encountered yet complex guidance scenario, i.e., mixed manned and unmanned vessels, is discussed in detail. In such cases, the International Regulations for Preventing Collisions at Sea (COLREGs) need to be complied during the autonomous guidance process. Leveraging the existing USV guidance methods, a velocity obstacle (VO) based approach is proposed and demonstrated to show how the USV autonomous guidance can be achieved.

3.2 USV PATH PLANNING

Path planning is an important module for USVs' autonomy and is fundamental in accomplishing specific autonomous tasks. Generally, USV path planning methods can be divided into classical path planning and intelligent path planning.

Depending on whether the environment information is known a priori, the classical path planning for USVs can be divided into global path planning [2] and local path planning [3]. Global path planning mainly deals with static environments, assuming that the complete model of the environment is known. Then, a safe path

DOI: 10.1201/9781003660590-3

from the origin to the destination can be found. Typical global path planning algorithms are Dijkstra [4], A-star (A*) [5], Rapidly-exploring Random Tree (RRT) [6], etc. These global path planning algorithms are implemented offline and could be time-consuming. Local path planning emphasizes on avoiding dynamic obstacles and generating local collision-free paths. Due to the online computations, local path planning algorithms generally need to be computationally efficient. The artificial potential field (APF) method [7] based on the attraction and repulsion mechanisms is a local path planning algorithm, and has the risk of falling into the local minima. The vector field histogram algorithm is proposed and used to deal with local path planning problems with obstacles near target points [8]. However, the vector field method is difficult to achieve satisfactory robustness in complex environments. The velocity space methods such as the dynamic window approach [9] are also used to deal with these local path planning problems that require collision avoidance online. This kind of methods are adaptive to the real and complex environments, and can avoid the possible collisions in the velocity space by selecting the feasible velocities within dynamic constraints [10].

In addition, global path planning and local path planning can be combined to solve the problems of path planning for USVs. For example, in [11], a dynamics-constrained global-local hybrid path planning scheme incorporating the global path planning and local hierarchical architecture is proposed for the USV. Besides the USV dynamics constraints, the marine traffic rules especially in situations with mixed manned and unmanned marine vehicles are necessarily complied with. COLREGs have been considered in various global and local path planning algorithms for USVs [12, 13, 10]. In [14], a goal-biased RRT approach for the rapid path planning in marine environments is proposed. In this method, an potential field inspired steering method is adopted to further smooth the path. The optimal collision-free path planning problem for multi-robots is considered in [15]. A model predictive APF method is proposed to solve the collision avoidance planning problem considering COLREGs and maneuverability rules [16]. Besides, the A* algorithm can be combined with an improved APF algorithm to achieve the global and local path planning tasks. The optimal control based methods are also widely used in both global and local path planning problems [17, 18]. Usually, problem specific designs will be combined with the classical optimal control methods.

In recent years, intelligent path planning methods have attracted increasing attention. The intelligent path planning methods include mainly the evolutionary algorithm based path planning [19] and Reinforcement Learning (RL) based path planning [20]. Different with the classical accurate path planning algorithms, most of the intelligent methods can be adaptive to the complex, dynamic and unknown scenarios, without requiring smoothness and differentiability of the problem. The evolutional algorithms inspired by biological behaviors, such as the Genetic Algorithm (GA) [21], Particle Swarm Optimization (PSO) [22], Ant Colony Optimization (ACO) [23], have been widely utilized for USV path planning. However, evolution-based algorithms cannot guarantee the global optimal solution. Solutions close to the optimal solution [17] can only be provided. The RL algorithms can obtain the best strategy through the interaction with the environment [20]. There has been a growing

TABLE 3.1 An overview of different path planning methods for USVs.

	Types	Algorithms	Main features
Classical methods	Global	Dijkstra [4]	1) Prior known environment; 2) static environment; 3) off-line; 4) time-consuming.
		A* [5]	
		RRT [6]	
	Local	Potential field [7]	1) Dynamic obstacle avoidance; 2) real-time; 3) local minima.
		Vector field [8]	
		Velocity space [9][10]	
		Optimal path planning [17][18]	
	Hybrid	Combinations of multiple algorithms [11][15]	1) Hierarchical path planning and improvement; 2) complexity.
Intelligent methods	Evolution algorithms	GA [21]	1) Biological behavior inspired; 2) complete but not optimal.
		PSO [22]	
		ACO [23]	
	RL	Q-learning [20][24][25]	1) Environment-interaction; 2) complex and dynamic scenarios; 3) black-box mode; 4) time-consuming model training.
		deep RL [26][27]	

interest in the path planning methods based on RL [24]. For example, a path planning method based on Q-learning, which is a classic RL algorithm, is proposed for the autonomous path planning of USV [25]. Combined with the deep neural network, the deep Q-learning algorithm is utilized in an elimination collision method for USVs [26]. Similarly, the deep RL algorithm is also utilized for the path planning of multiple USVs [27]. In complex situations, the RL methods might show superior performance than the traditional methods. However, the main issue concerned with the RL models are the black-box nature and the potential data issue.

Overall, the different path planning methods for USVs are concluded and classified in Table 3.1.

3.3 USV GUIDANCE LAWS

With the planned geometric path, guidance laws can be designed to guide the USV to approach to and finally maneuver along the path. The guidance laws are frequently integrated with the tracking control problems.

For guiding the USV towards the planned path, there are several commonly used techniques, e.g., the Pure Pursuit (PP) guidance [28], Line Of Sight (LOS) guidance [29, 30, 31] and Constant Bearing (CB) guidance [32]. Unlike the PP and CB guidance methods, the LOS guidance is a three-point guidance method as it involves a reference point in addition to the USV and the virtual target [29]. For USVs, the impact of ocean currents and drift angles are usually considered in the variants of the conventional LOS, e.g., the extended state observer-based LOS guidance [33], composite LOS guidance [34], error-constrained LOS guidance [35], etc. However, the conventional LOS guidance cannot be used directly in tracking curved paths. Therefore, the prospective distance has been adapted [36, 37] based on the path information or USV velocity information. Specifically, the change rate of the lookahead distance is adapted according to the USV speed in [36], and the path curvature information is considered in the lookahead distance in [37].

Compared to LOS, PP guidance is more suitable for tracking curved paths. It does not need to design complex lookahead distance adjustments and is more sensitive to dynamic virtual targets [38]. However, compared to LOS, the controller parameter tuning is more stringent due to overshoot issues [39]. The CB guidance is widely used in the missile community, and has the same engagement geometry as the PP guidance when used for USVs [40]. Since the guidance vectors of CB are composed of the velocity vectors of the follower vehicle and the virtual target, it is particularly useful in scenarios where the target's position is constantly changing. By adapting and adjusting the vehicle's heading, it is ensured that the vehicle remains on a direct course with the target's position [41]. Similarly, the CB guidance also faces potential overshoot issues [40].

3.4 COLREGS-COMPLIANT PATH PLANNING OF USVS

With the emergence of intelligent ships with different autonomy levels, the ship encountering scenarios can be roughly categorized into three types, i.e.,

- The encountering of traditional ships;

- The encountering between traditional and intelligent ships;

- The encountering of intelligent ships.

For the third scenario, the overall safety and optimality can usually be achieved via the cooperation among the involved ships, not necessarily following certain navigation rules [42]. However, for the other two scenarios, since the traditional ships generally navigate following COLREGs [43], regulation aware path planning methods need to be developed for ships that encounter with them.

However, most of the works mentioned above only focus on the efficiency and feasibility of the path, and have not considered path planning in conjunction with the navigation rules. Methods have also been proposed to improve the above works so that the ship maneuvers obeying COLREGs, such as fuzzy logic [59], interval programming [60], and local normal distribution-based trajectory algorithm [61]. However, these methods are difficult to be applied to the multi-ship encountering scenarios where multiple rules need to be satisfied simultaneously. A modified APF method is proposed in [62] to handle the scenario involving multiple-ships. COLREGs-constrained behaviors by adding appropriate functional and safety requirements in the corresponding virtual forces are achieved. Course alteration and speed changing in compliance with COLREGs can also be achieved by the linear extension algorithm [63]. However, the course alteration maneuver should in general be preferred over speed changes according to an experienced seamanship's practice. To achieve this, the multi-objective optimization approach proposed in [64] generates both experience and COLREGs-compliant paths. Similarly, the ship collision avoidance approach based on model predictive control [65] meets the requirements of COLREGs by constraining the optional control behaviors. However, in the complex marine environment, the acquired information that is used by the decision system is often uncertain and noisy. This brings further challenges to the path planning problem.

Although COLREGs set rules on how to maneuver in general for avoiding ship collisions, they do not provide quantitative criteria regarding when to activate the procedure and how to maneuver at the operational level. At the same time, in multi-ship encountering scenarios, the collision risks with multi-target ships are different. In most studies, the real-time ship collision risk is usually monitored by calculating the closest point of approach (CPA), such as the distance to the closest point of approach (DCPA), and/or the time to the closest point of approach (TCPA) [66], [67]. Other approaches set the domain of ship safety, and define the collision risk through the conflict between ships [68], [69]. Based on the ship domain, using the concept of the probability flow, the collision risk between two moving intelligent ships are evaluated considering the uncertainty in the ship trajectory [70]. Considering the uncertainty in the collision risk, a funnel library method is proposed [71], which enables the robot to perform real-time planning in an environment with complex geometric constraints. Fuzzy logic methods are also widely used in the risk assessment between ships, extending the rule-based approach by taking into account different influence parameters as the input membership functions of risks [72], [73]. However, most of the studies do not explicitly and effectively consider the dynamic collision risk changes between the own ship and the target ships.

Next, an advanced regulation aware dynamic path planning approach is proposed for USVs. The collision risk, the uncertainties in the dynamic environment, and the multi-target ships in compliance with COLREGs in complex encountering scenarios are considered uniformly. This guarantees the safety of the collision avoidance path planning of USVs in practical applications.

3.5 COLLISION RISK ASSESSMENT IN MULTI-SHIP SCENARIO

It is necessary to evaluate the dynamic collision risk of the own ship and the target ships in real time in order to determine how to plan the path. Five parameters, namely the TCPA, DCPA, relative distance, relative bearing, and ship speed ratio, are considered as the most relevant in determining the collision risks. Then, fuzzy membership functions are proposed to calculate the collision risk indices (CRI) between the own ship and all the encountered target ships. The maximum CRI value is compared with a safety CRI threshold to serve as the activation condition in the regulation aware dynamic path planning method.

Let the subscript o denote the own ship and $i = 1, 2, \cdots, n$ denote the n target ships. Let $\boldsymbol{p}_o = [x_o, y_o]^{\mathrm{T}}$, $\boldsymbol{p}_i = [x_i, y_i]^{\mathrm{T}}$ and $\boldsymbol{v}_o = [v_o^x, v_o^y]^{\mathrm{T}}$, $\boldsymbol{v}_i = [v_i^x, v_i^y]^{\mathrm{T}}$ denote the position and velocity vectors of the own ship and the target ships over a 2-D marine surface, respectively. Then, the CPAs between the own ship and the target ships are computed as follows:

$$T_{\text{CPA}}^i = \frac{(\boldsymbol{p}_o - \boldsymbol{p}_i)^{\text{T}} (\boldsymbol{v}_o - \boldsymbol{v}_i)}{\|\boldsymbol{v}_o - \boldsymbol{v}_i\|^2}, \tag{3.1}$$

$$D_{\text{CPA}}^i = \|(\boldsymbol{p}_o + \boldsymbol{v}_o T_{\text{CPA}}^i) - (\boldsymbol{p}_i + \boldsymbol{v}_i T_{\text{CPA}}^i)\|, \tag{3.2}$$

where $\|\cdot\|$ is the Euclidean norm. The risk membership function of D_{CPA}^i is calculated as:

$$R_{D_{\text{CPA}}}^i = \begin{cases} 1, & D_{\text{CPA}}^i < D_0 \\ \frac{1}{2}(1 - \sin \frac{\pi}{D_1 - D_0}(D_{\text{CPA}}^i - \frac{D_1 + D_0}{2})), & D_0 \le D_{\text{CPA}}^i < D_1 \\ 0, & D_{\text{CPA}}^i \ge D_1 \end{cases} \tag{3.3}$$

where D_0 is the minimum safety distance to the closest point of approach, and D_1 is the enough safety distance to the closest point of approach. The D_{CPA}^i between ships has a great influence on the risk of ship collisions. According to (3.3), the larger the D_{CPA}^i is, the less dangerous the other ship is to the own ship. According to [72], and the collision cone characteristics of the VO algorithm, D_0 is set as the sum of the radii of the own ship and the target ship, and $D_1 = 4D_0$.

The risk membership function of T_{CPA}^i is then calculated as:

$$R_{T_{\text{CPA}}}^i = \begin{cases} 1, & T_{\text{CPA}}^i < T_0 \\ \frac{1}{2}(1 - \sin \frac{\pi}{T_1 - T_0}(T_{\text{CPA}}^i - \frac{T_1 + T_0}{2})), & T_0 \le T_{\text{CPA}}^i < T_1 \\ 0, & T_{\text{CPA}}^i \ge T_1 \end{cases} \tag{3.4}$$

where T_0 is the minimum safety time to the closest point of approach, and T_1 is the enough safety time to the closest point of approach. The T_{CPA}^i between ships has a great influence on the risk of ship collisions. According to (3.4), the larger the T_{CPA}^i is, the less dangerous the other ship is to the own ship. The parameters T_0 and T_1 can be calculated according to S_0 and S_1.

$$T_0 = \begin{cases} \frac{\sqrt{S_0^2 - D_{\text{CPA}}^i{}^2}}{\|\boldsymbol{v}_o - \boldsymbol{v}_i\|}, & D_{\text{CPA}}^i <= S_0 \\ \frac{S_0 - D_{\text{CPA}}^i}{\|\boldsymbol{v}_o - \boldsymbol{v}_i\|}, & D_{\text{CPA}}^i > S_0 \end{cases} \tag{3.5}$$

$$T_1 = \begin{cases} \frac{\sqrt{S_1^2 - D_{\text{CPA}}^i{}^2}}{\|\boldsymbol{v}_o - \boldsymbol{v}_i\|}, & D_{\text{CPA}}^i <= S_1 \\ \frac{S_1 - D_{\text{CPA}}^i}{\|\boldsymbol{v}_o - \boldsymbol{v}_i\|}, & D_{\text{CPA}}^i > S_1 \end{cases} \tag{3.6}$$

where S_0 is set as four times of the sum of the radii of the own ship and the target ship, and S_1 is calculated as:

$$S_1 = 1.7 \cos(B^i - 19°) + \sqrt{4.4 + 2.89 \cos^2(B^i - 19°)}, \tag{3.7}$$

In some scenarios, the ship collision risk can be effectively evaluated using the CPA method. However, in a complex multi-ship encountering scenario with consideration of COLREGs, it is unreliable to only use the CPA value for the pre-collision check.

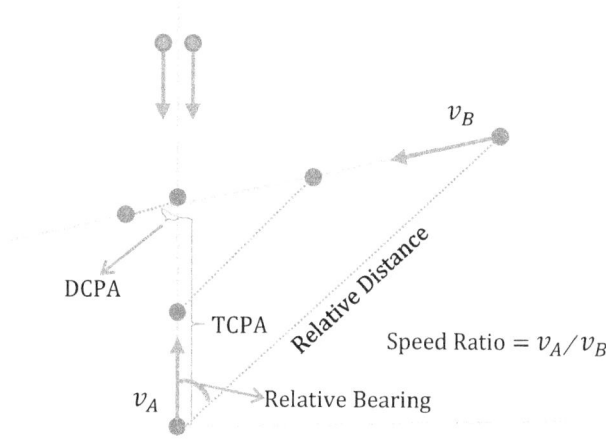

Figure 3.1: Dynamic collision risk assessment for ships based on the fuzzy theory, taking into account TCPA, DCPA, relative distance, relative bearing and ship speed ratio. The closest distance marked with a red line and blue dashed line indicates the relative distance between the own ship and the target ship at a certain time.

For example, for a head-on situation as shown in Fig. 3.1, the own ship and the target ship move towards each other. The target ship located on the right of the centerline of the own ship is more dangerous than the target ship on the left, which requires taking the relative bearing into account in calculating the collision risks. Generally, the situation of cross encountering is more risky and dangerous than that of the head-on encountering. Therefore, besides the conventional TCPA and DCPA, the relative distance, relative bearing and ship speed ratio are also considered in calculating the collision risks.

Specifically, the risk membership function of the relative distance is calculated as:

$$R_S^i = \begin{cases} 1, & S^i < S_0 \\ \frac{1}{2}(1 - \sin\frac{\pi}{S_1 - S_0}(S^i - \frac{S_1 + S_0}{2})) & , S_0 \leq S^i < S_1 \\ 0, & S^i \geq S_1 \end{cases} \tag{3.8}$$

where S^i is the relative distance between the own ship and the target ship i. According to (3.8), the closer the relative positions of the two ships are, the higher the risk of collision between the ships.

The target ships in different directions pose different degrees of danger to the own ship. According to the experience of collision avoidance [78], when the incoming ship is located at about 19°, the collision risk of the own ship is the greatest. When the incoming ship is located at about 199°, the collision risk is the lowest. Therefore, the membership function of the relative bearing of the target ship is established as follows:

$$R_B^i = \frac{1}{2}(\cos(B^i - 19°) + \sqrt{\frac{440}{289} + \cos^2(B^i - 19°)}) - \frac{5}{17}, \tag{3.9}$$

In general, the risk of the starboard side is larger than that of the port side, and the risk of the front side is larger than that of the rear side.

In a similar way, the membership function of the ship speed ratio is established as follows:

$$R_K^i = \frac{1}{(1 + (\frac{K_0}{K^i})^2)}, \tag{3.10}$$

where the ship speed ratio is defined as $K^i = \frac{V_o}{V_i}$, and K_0 is generally set as 1. When $K^i = 1$, the two ships have the same speed, and $R_K^i = 0.5$ is the boundary between safety and danger.

Then, for a specific target ship i, we are able to define the collision risk vector as

$$R^i = [R_{T_{\text{CPA}}}^i, R_{D_{\text{CPA}}}^i, R_S^i, R_B^i, R_K^i]^{\text{T}}. \tag{3.11}$$

To balance the influence of each of the five factors, define the weight vector

$$W = [W_{T_{\text{CPA}}}, W_{D_{\text{CPA}}}, W_S, W_B, W_K]^{\text{T}}, \tag{3.12}$$

where $W_{T_{\text{CPA}}} + W_{D_{\text{CPA}}} + W_S + W_B + W_K = 1$. Then, the CRI value for the target ship i is calculated as:

$$CRI^i = W^{\text{T}} R^i, \tag{3.13}$$

which indicates that the range of the CRI values are all between [0-1]. In the multi-ship encountering scenario, the overall collision risk is determined by the most dangerous target ship, i.e.,

$$CRI = \max\{CRI^i\}, \qquad i = 1, \ldots, n \tag{3.14}$$

This CRI value is then compared with a safety threshold to activate the collision avoidance maneuvers, as to be detailed in the next. In addition, the risk weight vectors and the CRI minimum safety threshold are set in a way [72] that the weighted sum of TCPA and DCPA is greater than the CRI threshold when the situation is at risk. Therefore, obstacle avoidance maneuvers can be triggered when collision risks are encountered.

3.6 REGULATION AWARE DYNAMIC PATH PLANNING

This section proposes a VO-based method for the collision-free path planning in the multi-ship encountering scenario when collision risks are above the safety threshold. The VO method generates a conical obstacle in the velocity space for the dynamic collision avoidance based on the velocities of multiple moving ships. Moreover, considering the ubiquitous uncertainties in the measured information, the collision cone in the conventional VO is extended. The uncertain VO path planning algorithm is proposed using a robust tube idea. Under a certain degree of confidence, the probabilistic impact is considered in a unified manner to achieve robust collision-free path planning in compliance with COLREGs.

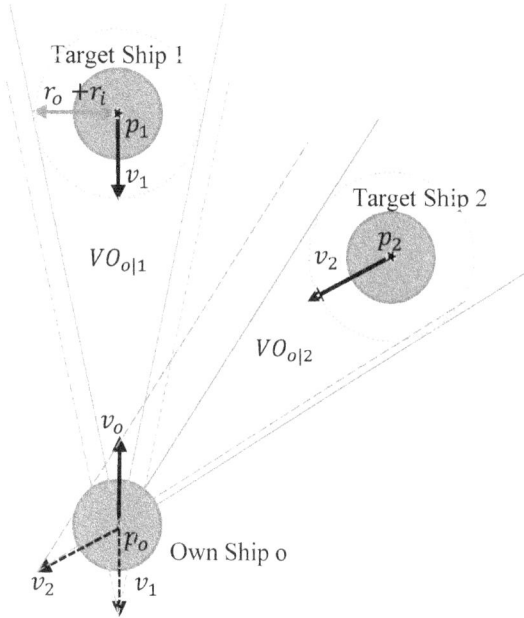

Figure 3.2: Velocity obstacle in the multi-ship encountering scenario.

3.6.1 Collision-free path planning with VO

The basic idea of VO is to turn a general dynamic obstacle avoidance problem to a static velocity reachability problem by defining a cone shaped obstacle in the velocity space. The multi-ship encountering scenario in the velocity space is shown in Fig. 3.2. Specifically, head-on (target ship 1) and crossing from the right (target ship 2) encountering situations are observed. Both the own ship and the target ships are defined as discs. Without loss of generality, the discs' diameters are set as the ships' lengths.

In Fig. 3.2, $VO_{o|i}, i = 1, 2$ is the velocity set that causes the collision between the own ship o and a target ship i. Therefore, the condition that the own ship avoids the collision with the target ship i is $v_o \notin VO_{o|i}$. For each target ship, the radius of the blue dotted circle is the sum of the radii of the own ship and target ship. The solid lines form the collision cones for the relative velocities between the own ship and the target ships. The dotted lines form the collision cones for the absolute velocities by translating the cones considering the target ship velocities. Therefore, it leads to a collision when the velocity v_o is inside the dash-line cones, or equivalently when the relative velocity $v_o - v_i$ is inside the solid-line cones.

Mathematically, let $\lambda(\boldsymbol{p}, \boldsymbol{v}) = \{\boldsymbol{p} + t\boldsymbol{v} | t \geq 0\}$ be a ray starting from $\boldsymbol{p}$ going into the direction of $\boldsymbol{v}$. Define the Minkowski sum $\oplus$ as $A \oplus B = \{a + b | a \in A, b \in B\}$, and $O \oplus I$ denote Minkowski sum of the discs of the own ship and the target ships. Define $-O = \{-o | o \in O\}$ as the reflection. Then,

$$VO_{o|i} = \{\boldsymbol{v}_o | \lambda(\boldsymbol{p}_o, \boldsymbol{v}_o - \boldsymbol{v}_i) \cap I \oplus -O \neq \emptyset\} \tag{3.15}$$

is the set of velocity obstacle solutions of the own ship in the multi-ship encountering scenario. Intuitively, (3.15) means that if the own ship sails at a speed $\boldsymbol{v}_o \in VO_{o|i}$, a collision will occur at a future moment.

Since both the own ship and the target ships are simplified as discs, (3.15) can be simplified as

$$VO_{o|i} = \{\boldsymbol{v}_o | \exists t > 0 :: (\boldsymbol{v}_o - \boldsymbol{v}_i)t \in D(\boldsymbol{p}_i, r_o + r_i)\}, \qquad (3.16)$$

which origins from $\boldsymbol{p}_o$, and $D(\boldsymbol{p}_i, r_o + r_i)$ is an open disc of radius $r_o + r_i$ centered at $\boldsymbol{p}_i$; r_o and r_i are the radii of the own ship and the target ships, respectively. For all the target ships $i = 1, \cdots, n$, the velocity obstacle set in the multi-ship encountering scenario is $\bigcup\limits_{i=1}^{n} VO_{o|i}$. The own ship must sail at the velocity that is not within this set to guarantee safety.

3.6.2 Robust collision-free path planning with uncertain VO

From (3.16), the velocity and position of the target ships are necessary to be known to calculate the velocity obstacles. However, in complex marine environment, these measured information is usually uncertain and probabilistic [79], [80]. On the one hand, the information measured by the onboard sensors, such as lidar and cameras, could have noises; on the other hand, the target ship motions could be affected by the environmental disturbances such as wind, waves, and currents. The uncertainties in the velocities and positions of the target ships could lead to infeasibility of the algorithm, and thus raise serious safety issues in the multi-ship encountering scenarios. Therefore, we next propose a robust collision-free path planning method based on uncertain VO.

It is assumed that there exist uncertainties that follow Gaussian distributions in the target ships' radii, velocities, and positions. Specifically, $r_i \sim N(\hat{r}_i, \Sigma_{r_i})$, $\boldsymbol{v}_i \sim N(\hat{\boldsymbol{v}}_i, \Sigma_{\boldsymbol{v}_i})$, and $\boldsymbol{p}_i \sim N(\hat{\boldsymbol{p}}_i, \Sigma_{\boldsymbol{p}_i})$, where $\hat{r}_i$, $\hat{\boldsymbol{v}}_i$, and $\hat{\boldsymbol{p}}_i$ are the corresponding mean values; Σ_{r_i}, $\Sigma_{\boldsymbol{v}_i} = \mathrm{diag}([\delta_{\boldsymbol{v}_i}^{x\,2}, \delta_{\boldsymbol{v}_i}^{y\,2}])$, $\Sigma_{\boldsymbol{p}_i} = \mathrm{diag}([\delta_{\boldsymbol{p}_i}^{x\,2}, \delta_{\boldsymbol{p}_i}^{y\,2}])$ are the corresponding variance or covariance matrices, and δ. means the corresponding standard values; diag means the diagonal matrices. Then, the deterministic VO in (3.16) is extended to

$$UVO_{o|i} = \{\boldsymbol{v}_o | \exists t > 0 :: V_{\mathrm{UVO}}^i t \in D(P_{\mathrm{UVO}}^i, R_{\mathrm{UVO}}^i)\}, \qquad (3.17)$$

where

$$R_{\mathrm{UVO}}^i = r_o + r_i \sim N(r_o + \hat{r}_i, \Sigma_{r_i}), \qquad (3.18)$$

$$V_{\mathrm{UVO}}^i = \boldsymbol{v}_o - \boldsymbol{v}_i \sim N(\boldsymbol{v}_o - \hat{\boldsymbol{v}}_i, \Sigma_{\boldsymbol{v}_i}), \qquad (3.19)$$

$$P_{\mathrm{UVO}}^i = \boldsymbol{p}_i \sim N(\hat{\boldsymbol{p}}_i, \Sigma_{\boldsymbol{p}_i}). \qquad (3.20)$$

Since these stochastic uncertainties are with infinite support, it is impossible and also unnecessary to consider those extreme conditions. Therefore, we only consider the collision avoidance with a certain probability p_r. As shown in Fig. 3.3, for each target ship, the small yellow dotted circle is considered as the position uncertainty of the target ship, and the bigger yellow dotted circle is considered as

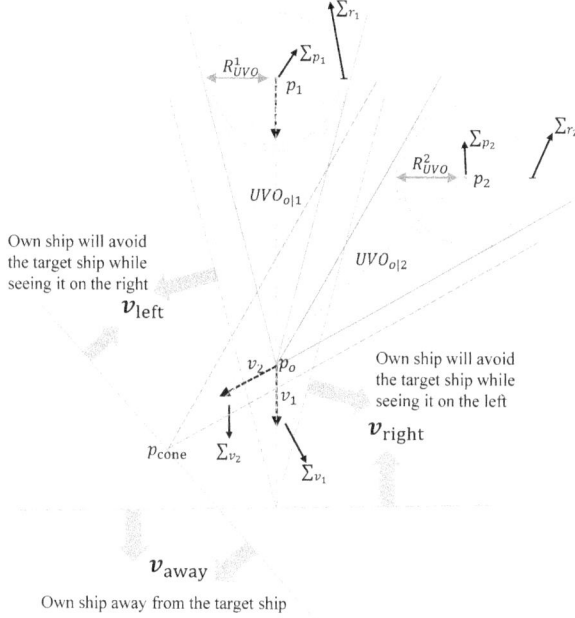

Figure 3.3: Construction of the COLREGs with the uncertain velocity obstacle. Among them, the orange collision cones are obtained by considering the uncertainty of velocity, position, and radius, and $\mathcal{V}_{\text{left}}$, $\mathcal{V}_{\text{right}}$, $\mathcal{V}_{\text{away}}$, are different velocity sets that conform to the COLREGs. Taking the velocity in $\mathcal{V}_{\text{away}}$, the own ship will see the target ships passing on their left. This operation applies to head-on and other encountering situations as well.

the velocity uncertainty. The bigger blue dotted circle is the Minkowski sum of a disc D_{cone}, where D_{cone} represents the extended circle of the collision cone. Specifically, $r_i = \hat{r}_i + \Delta r_i$, $\boldsymbol{v}_i = \hat{\boldsymbol{v}}_i + \Delta \boldsymbol{v}_i$, $\boldsymbol{p}_i = \hat{\boldsymbol{p}}_i + \Delta \boldsymbol{p}_i$, where $\Delta r_i = \sqrt{2}\delta_r \text{erf}^{-1}(p_r)$, $\Delta \boldsymbol{p}_i = [\Delta \boldsymbol{p}_i^x, \Delta \boldsymbol{p}_i^y]^T = [\sqrt{2}\delta_{\boldsymbol{p}_i^x} \text{erf}^{-1}(p_r), \sqrt{2}\delta_{\boldsymbol{p}_i^y} \text{erf}^{-1}(p_r)]^T$, $\Delta \boldsymbol{v}_i = [\Delta \boldsymbol{v}_i^x, \Delta \boldsymbol{v}_i^y]^T = [\sqrt{2}\delta_{\boldsymbol{v}_i^x} \text{erf}^{-1}(p_r), \sqrt{2}\delta_{\boldsymbol{v}_i^y} \text{erf}^{-1}(p_r)]^T$. Therefore, D_{cone} could be expressed as:

$$D_{\text{cone}} = D(\hat{\boldsymbol{p}}_i, \|\Delta \boldsymbol{p}_i\|) \oplus (r_o + \hat{r}_i + \Delta r_i). \tag{3.21}$$

$D(\hat{\boldsymbol{p}}_i, \|\Delta \boldsymbol{p}_i\|)$ is an open disc of radius $\|\Delta \boldsymbol{p}_i\|$ centered at $\hat{\boldsymbol{p}}_i$.

The solid lines form the collision cones considering the uncertainties between the own ship and the target ships. The collision cones are expressed as $C(\boldsymbol{p}_o, D_{\text{cone}})$. Where $\boldsymbol{p}_o$ is the cone point and D_{cone} is the extended uncertainty disc. Define the linear transformation of $\boldsymbol{p}_{\text{cone}}$ as $\boldsymbol{p}_o$. Dotted lines of the cones $C(\boldsymbol{p}_{\text{cone}}, D_{\text{cone}})$ are that moving the edges of the uncertainty collision cones $C(\boldsymbol{p}_o, D_{\text{cone}})$ out perpendicularly by an amount large enough, encompassing the linear transformation of a disc by the covariance matrix $\Sigma_{\boldsymbol{v}_i}$, and $\boldsymbol{p}_{\text{cone}}$ is the cone point of the $UVO_{o|i}$. This is satisfied by assuming that robustness is only against a certain probability p_r of the infinite support stochastic uncertainties. We consider 2 standard deviation around the mean.

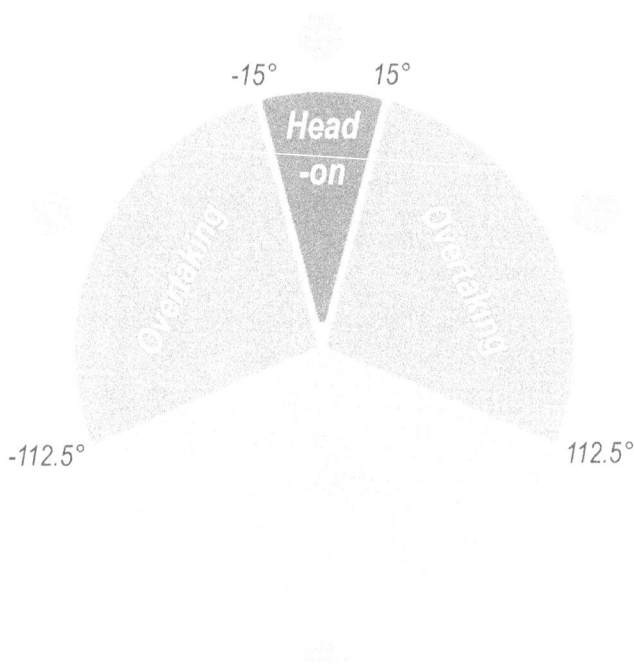

Figure 3.4: The main encounter situations between the intelligent ship and target ships. The center of the circle is the own ship, and the target ships come from all directions.

An uncertain VO can then be designed to be robust to 95% of the measurement uncertainties.

The uncertain VO could effectively deal with the existing uncertainty. It leads to a collision when the velocity $\boldsymbol{v}_o$ is inside the dash-line cones, that is, $\boldsymbol{v}_o \in UVO_{o|i}$, or equivalently when the relative velocity $\boldsymbol{v}_o - \boldsymbol{v}_i$ is inside the solid-line cones. For all the target ships $i = 1, \cdots, n$, the uncertain velocity obstacle set in the multi-ship encountering scenario is $\bigcup_{i=1}^{n} UVO_{o|i}$. In this condition, the own ship sailing with a velocity that is not within this unioned set will guarantee the navigation safety, even if there exist uncertainties.

3.6.3 Regulation aware robust collision-free path planning

It is necessary to consider the environment shared by USVs and traditional ships over the water. This requires that the USVs comply with COLREGs and are able to plan path safely and avoid collisions with other USVs or traditional ships. Three major situations are considered with COLREGs, i.e., overtaking, crossing, and head-on, as shown in Figs. 3.4 and 3.5. The relevant rules are [43], see Appendix for more details.

There are different views on the definition of the degrees of encountering scenarios. For the head-on encountering scenarios of traditional ships, the angle is generally defined as 5-6 degrees left and right in front of the ship. In this book, considering the

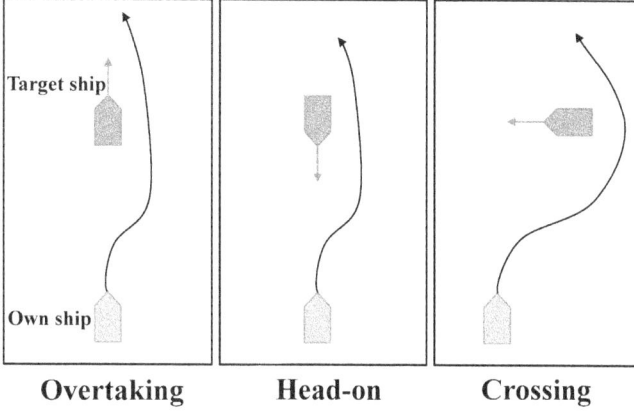

Overtaking　　　　　**Head-on**　　　　　**Crossing**

Figure 3.5: For rule 13 to 17, compliance requirements given by the COLREGs. From left to right: (a)Overtaking, (b)head-on and (c)crossing situations. The black arrows indicate compliance actions of COLREGs in each situation.

perception distance and accuracy limitations of the USVs, the angle is expanded to 15 degrees left and right in front of the own ship, which is more in line with the actual situation. Based on the previous section, we combine the uncertain VO algorithm and COLREGs to propose the regulation aware dynamic collision-free path planning method.

Each velocity reachable set of the target ship is divided into three subsets according to COLREGs, namely $\mathcal{V}_{\text{left}}^i$, $\mathcal{V}_{\text{right}}^i$, $\mathcal{V}_{\text{away}}^i$. The division of the three-part subsets is obtained by the angular bisector tangent to the uncertainty speed obstacle cone $C(\boldsymbol{p}_{\text{cone}}, D_{\text{cone}})$. The subsets are perpendicular to the cone.

When the own ship and the target ship are in a head-on situation, COLREGs require both ships to turn right to avoid obstacles. Therefore, the own ship need to follow a velocity in $\mathcal{V}_{\text{right}}^i$, which is,

$$\mathcal{V}_{\text{right}}^i = \{\boldsymbol{v}_o | \boldsymbol{v} \notin UVO_{o|i}, ((\hat{\boldsymbol{p}}_i + \boldsymbol{p}_{\text{cone}} - \boldsymbol{p}_o) - \boldsymbol{p}_{\text{cone}})^{\text{T}} (\boldsymbol{v}_o - \boldsymbol{v}_i) > 0, \boldsymbol{v}_o \notin \mathcal{V}_{\text{left}}^i\}. \tag{3.22}$$

Then, the corresponding subsets of all the target ships are merged in a multi-ship encountering scenario as $\mathcal{V}_{\text{left}}$, $\mathcal{V}_{\text{right}}$ and $\mathcal{V}_{\text{away}}$, as shown in Fig. 3.3. These merged subsets are calculated as:

$$\{\mathcal{V}_{\text{left}}, \mathcal{V}_{\text{right}}, \mathcal{V}_{\text{away}}\} = \{\bigcap_{i=1}^{n} \mathcal{V}_{\text{left}}^i, \bigcap_{i=1}^{n} \mathcal{V}_{\text{right}}^i, \bigcap_{i=1}^{n} \mathcal{V}_{\text{away}}^i\}. \tag{3.23}$$

Rule 8 of COLREGs also recommends that course changes should be prioritized over speed changes if there is enough free space available, and that maneuvers must be made in ample time. To better comply with Rule 8, when the velocity magnitude or the direction of the reachable velocity set can be changed at the same time, we usually choose to change the velocity direction. Let α_i, β_i be defined as half of the cone angle of $C(\boldsymbol{p}_o, D_{\text{cone}})$ and $C(\boldsymbol{p}_{\text{cone}}, D_{\text{cone}})$, respectively, then

$$\alpha_i = \arcsin \frac{\|R^i_{\text{UVO}} + \|\Delta \boldsymbol{p}\|\|^2}{\|\hat{\boldsymbol{p}}_i - \boldsymbol{p}_o\|^2}, \tag{3.24}$$

$$\beta_i = \arccos \frac{((\hat{\boldsymbol{p}}_i + \boldsymbol{p}_{\text{cone}} - \boldsymbol{p}_o) - \boldsymbol{p}_{\text{cone}})^{\text{T}}(\boldsymbol{v}_o - \hat{\boldsymbol{v}}_i)}{\|((\hat{\boldsymbol{p}}_i + \boldsymbol{p}_{\text{cone}} - \boldsymbol{p}_o) - \boldsymbol{p}_{\text{cone}})\|^2 \|(\boldsymbol{v}_o - \hat{\boldsymbol{v}}_i)\|^2}. \tag{3.25}$$

Then, let $\alpha_i = \beta_i$ to calculate the new $\boldsymbol{v}_o$. When the collision risk value of multiple target ships exceeds the set threshold at the same time, it is necessary to select the maximum angle of β_i for avoidance. Another advantage of selecting the maximum angle is that the target ship can better perceive the movement intention of the own ship, so that the target ship could also take corresponding maneuvers in advance.

It should be noted that when the own ship acts as a stand-on vessel in various encountering situations, the own ship has no COLREGs constraints. Then, the target ship needs to take corresponding maneuvers to avoid collisions with the own ship. However, in the real encountering scene, the target ship could not take any measures, or the measures taken do not meet the requirements of COLREGs. Even if this happens, the own ship can still adopt the velocity in the reachable velocity set to safely avoid collisions with the target ship. Therefore, the proposed algorithm allows the own ship to safely avoid any moving obstacles obeying COLREGs.

3.7 SIMULATIONS AND DISCUSSIONS

In this section, we present simulation results to demonstrate the effectiveness of the proposed dynamic collision-free path planning method. Three situations, i.e., crossing, overtaking and head-on, in the marine environment are considered. The simulation experiments of multiple target ships are carried out, and the feasibility of the algorithm is verified with real AIS data.

3.7.1 Simulation settings

The proposed algorithm involves the uncertainties of the position, velocity and radius of the target ships. In the simulations, we set $\Sigma_{\boldsymbol{v}_i} = \text{diag}([0.01, 0.01])$, $\Sigma_{\boldsymbol{p}_i} = \text{diag}([0.01, 0.01])$, where diag means diagonal matrices. The relative safety distance between ships R_{sd} is 4.4. For the setting of collision risk assessment parameters, the minimum safety distance to the closest point of approach is $D_0 = 4$. The enough safety distance to the closest point of approach is $D_1 = 16$, and $S_0 = 16$. The weight vector $W = [0.32, 0.36, 0.14, 0.10, 0.08]^{\text{T}}$. Relevant parameters are summarized in Table 3.2.

The own ship evaluates the collision risk of each target ship in real time during the path planning process. When one of the target ships reaches the safety threshold, the own ship starts to perform the path planning and collision avoidance process according to the uncertain-VO algorithm. When the risk indices of multiple target ships exceed the set threshold simultaneously, the ships with the highest collision risk index need to be considered first. Then, according to the membership functions of the influencing factors, the collision risk evaluation matrix R of the target ship to the

TABLE 3.2 Parameter values in the simulation.

Parameter	Value
L	4
Σ_{r_i}	0.01
$\Sigma_{\boldsymbol{v}_i}$	$\mathrm{diag}([0.01, 0.01])$
$\Sigma_{\boldsymbol{p}_i}$	$\mathrm{diag}([0.01, 0.01])$
R_{safe}	0.6
R_{sd}	4.4
D_0	4
D_1	16
S_0	16
W	$[0.32, 0.36, 0.14, 0.10, 0.08]^{\mathrm{T}}$

own ship can be established. We consider 2 standard deviation around the mean. An uncertain VO can then be designed to be robust to $p_r = 95\%$ of the measurement uncertainties. In the simulations, all the algorithms are implemented in Python 3.8.8 on a platform with Intel(R) Core(TM) i5-10400 CPU @2.90 GHz.

3.7.2 Path planning with multiple target ships

This simulation shows the encountering situations between the own ship and multi-target ships under COLREGs. Considering the uncertainty in the radius, as well as position and speed of the three target ships, the size of uncertain-VO collision cone is expanded. Using the expanded collision cone, that is, under 95% confidence, the reliability of path planning is guaranteed.

Scenario 1: Overtake, Head-On. Fig. 3.6 shows the simulation process of the own ship and the target ships to verify the effectiveness of collision avoidance using the expanded uncertain-VO collision cone. The gray dotted line in the figure represents the left and right tangents of the expanded VO cone during the collision avoidance process. Target ships 1, 2, and 3 (green, yellow, and blue trajectories, respectively), and the own ship (red trajectories) use the uncertain-VO method to decide maneuvers. In Fig. 3.6(a), the own ship is located at [0,-50] m and navigates northward in a straight line at a speed of 1 m/s; two of the target ships are located at [0,-25] m and [0,-15] m, and both are navigating northward in a straight line at the speed of 0.5 m/s, and the third target ship is at a position of [5,45] m and is navigating at a speed of [0.05,-0.5] m/s. In the initial stage, for the target ship 1 and the target ship 2, the own ship is in the encountering scene of overtaking, and is in the head-on situation with the target ship 3.

Fig. 3.6(b) shows the own ship recognizes the overtaking situation and applies the COLREGS constraint. The collision risk of target ship 1 exceeds the set threshold, and the COLREGs constraints encourage the overtaking evasive maneuver to turn right. The real time collision risks between the own ship and target ships are shown in Fig. 3.7.

(a)

(b)

(c)

(d)

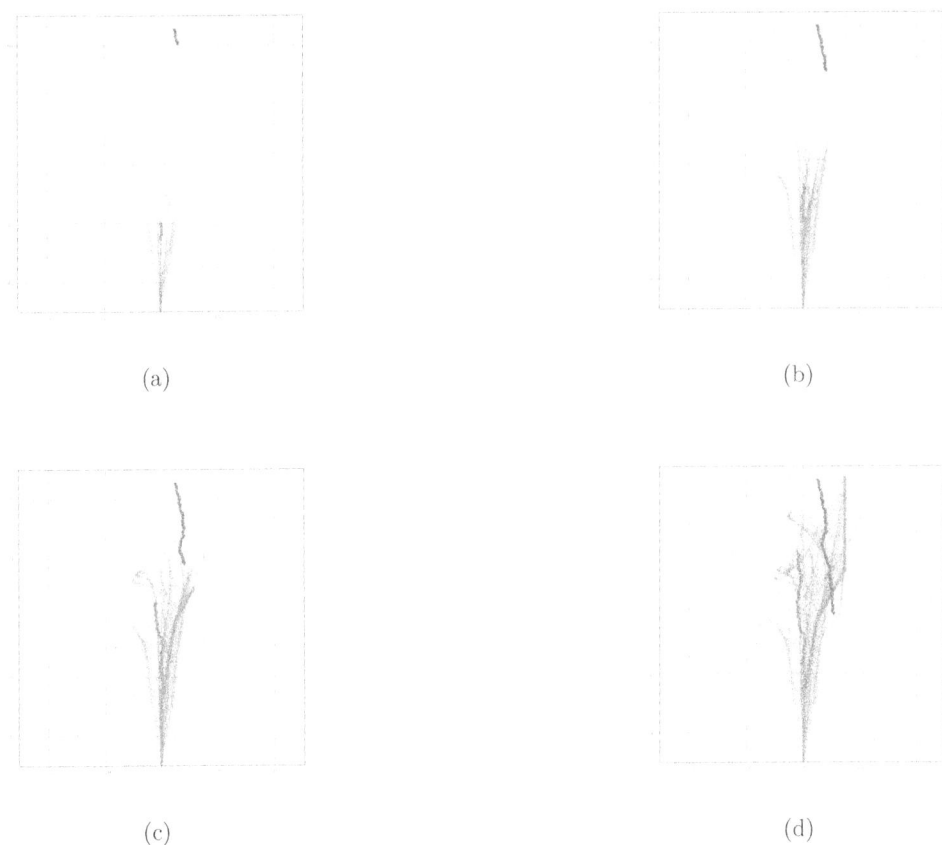

Figure 3.6: Scenario 1 of the encountering situation of multi-target ships under COL-REGs and uncertain-VO algorithm. Three target ships, with green, yellow, and blue trajectories, all exist uncertainties in the radius, velocity and position. (a) the initial stage of the four ships; (b) the maneuver in the overtaking situation; (c) the maneuver in the overtaking and head-on situations; (d) overall maneuvering path of the own ship and the target ships, respectively.

In Fig. 3.6(c), after the own ship has been moving for a period of time, target ship 3 has gradually approached, the own ship and target ship 1 have completed the collision avoidance process, and are still in the overtaking situation with target ship 2 and the head-on situation with target ship 3. At this time, the collision risk values of target ships 2 and 3 exceed the set threshold at the same time. Then, the maximum angle of β_i is selected for collision avoidance according to the COLREGs.

Fig. 3.6(d) shows successful collision avoidance with the target ships. When the collision risk value of each target ship is below the set threshold, the own ship navigates with initial velocity. Fig. 3.8(a) shows the distance between the own ship and the target ship by using the deterministic VO method regardless of the uncertainty of the target ship, where the red line is the set safety line. Fig. 3.8(b) shows the distance

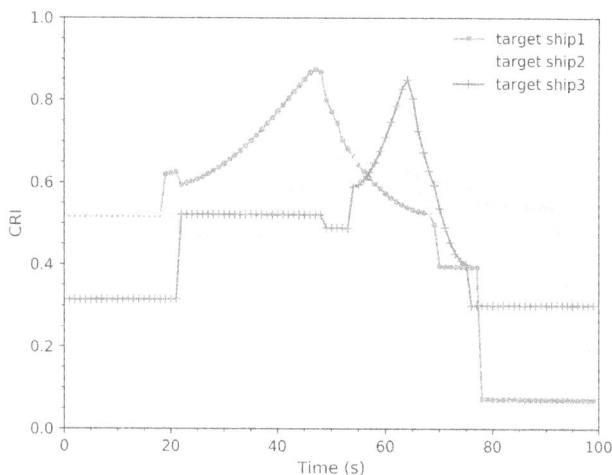

Figure 3.7: Real-time collision risk indices in scenario 1.

between the own ship and the target ship obtained by using uncertain-VO method considering the uncertainty of the three target ships.

Scenario 2: Overtake, Head-On, and Crossing. In this scenario, the own ship first overtakes the target ship 1 (green trajectory). As it maneuvers around it, other two targe ships (yellow, and blue trajectories) start to approach the own ship, forming a head-on and cross situation at the same time, as shown in Fig. 3.8

Fig. 3.9(a) shows the initial state of ships, and the own ship creates three collision cones. In Fig. 3.9(b), the CRI of the target ship 1 is greater than the safe CRI, and the own ship recognizes the overtaking situation. The own ship then takes a maneuver to starboard, which is in compliance with the COLREGs. The real time collision risks between the own ship and target ships are shown in Fig. 3.10.

In Fig. 3.9(c), in the process of avoiding target ship 1, the own ship detects the other target ships. Then, the own ship recognizes that it is in the overtake, head-on, and crossing situation at the same time. The CRI of target ship 1 and target ship 3 (blue trajectory) exceeds the set safety CRI. COLREGs constraints force the own ship to turn starboard to avoid the crossing target ship 3.

Fig. 3.9(d) shows that all the target ships are safely passed. Note that when the own ship avoid the target ship 3 successfully, the own ship is still in the overtaking situation with the target ship 1. In this moment, the target ship 1 is far enough and the CRI is relatively small. The own ship does not consider this situation to be overtaking. Eventually, the own ship initiates a turn maneuver with the initial velocity. Figs. 3.11(a) and 3.11(b) shows the distance between the own ship and the target ships by using the traditional VO or uncertain-VO method, respectively.

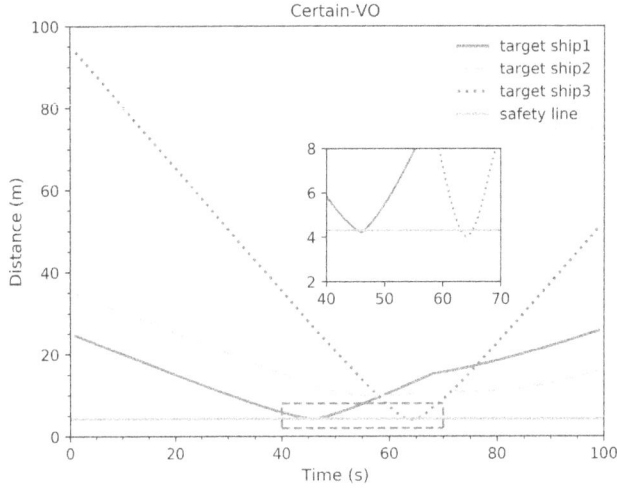

(a)

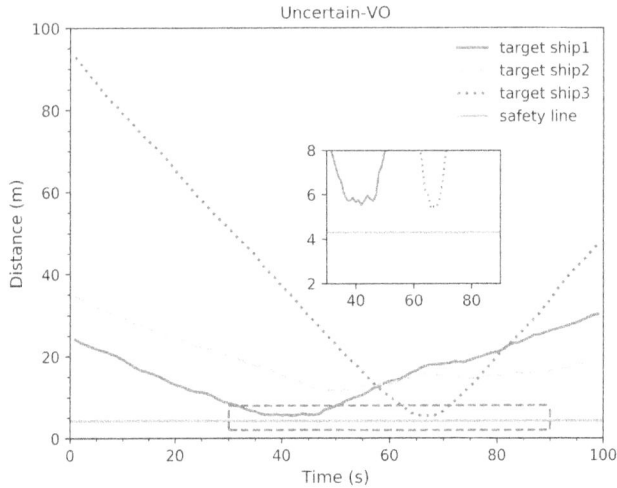

(b)

Figure 3.8: Relative distance of the intelligent ship path planning under COLREGs in scenario 1. (a) The distances between the own ship and the target ships without considering the uncertainties. (b) The distances between the own ship and the target ships using the uncertain-VO method considering the uncertainties of the target ships.

3.7.3 Verification with AIS data

According to the real automatic identification system (AIS) data in an area of the Yangtze River in Wuhan, China, effectiveness and the potential to be applied in practice of the proposed algorithm is further verified.

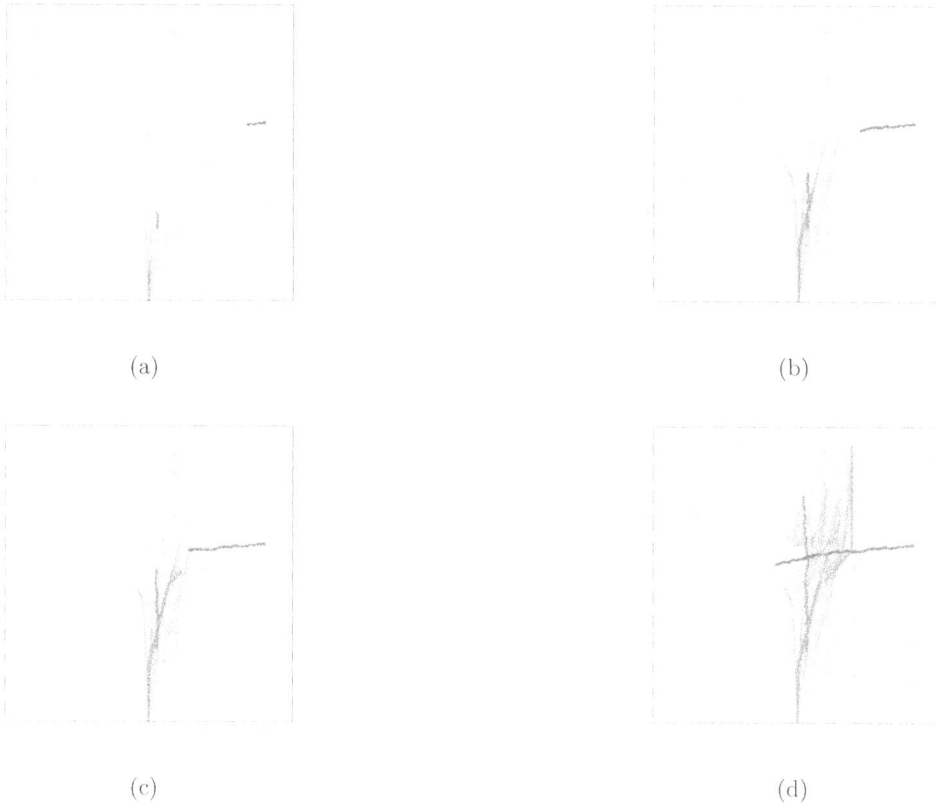

(a)

(b)

(c)

(d)

Figure 3.9: Scenario 2 of the encountering situation of multi-target ships under COL-REGs and uncertain-VO algorithm. (a) the initial stage of the four ships; (b) the maneuver in the overtaking situation; (c) the maneuver in the overtaking, crossing and head-on situations; (d) overall maneuvering path of the own ship and the target ships, respectively.

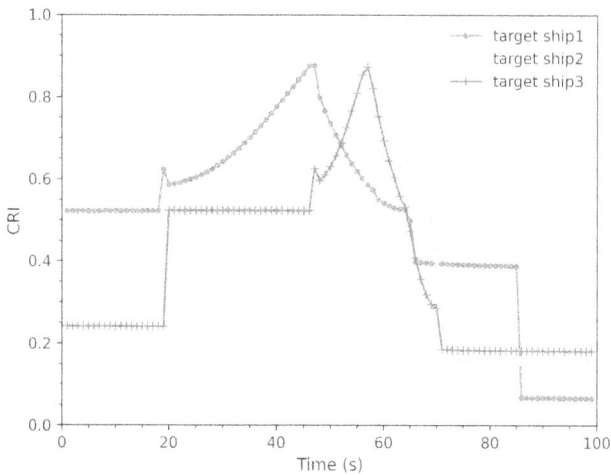

Figure 3.10: Real-time collision risk indices in scenario 2.

(a)

(b)

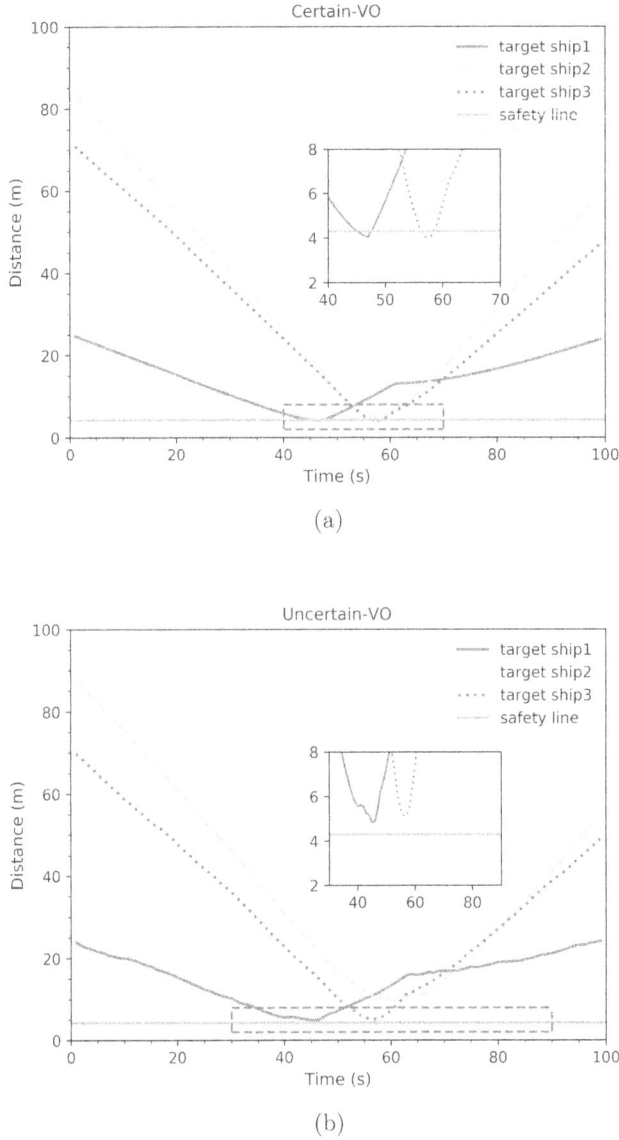

Figure 3.11: Relative distances of the intelligent ship path planning under COLREGs in scenario 2. (a) The distances between the own ship and the target ships without considering the uncertainties. (b) The distances between the own ship and the target ships using the uncertain-VO method considering the uncertainties of the target ships.

Fig. 3.12(a) shows the simulation at the beginning. The direction and length of the blue arrows represent the speed and direction of each ship, respectively. In Fig. 3.12(b), during the process, the own ship is in the situation of head-on, and detects the risk of collision with target ship 1. According to COLREGs, the own ship has the responsibility to turn right to avoid the collision. In the process of avoiding target ship 1, target ship 2 may also collide with the own ship, and is in a overtaking situation

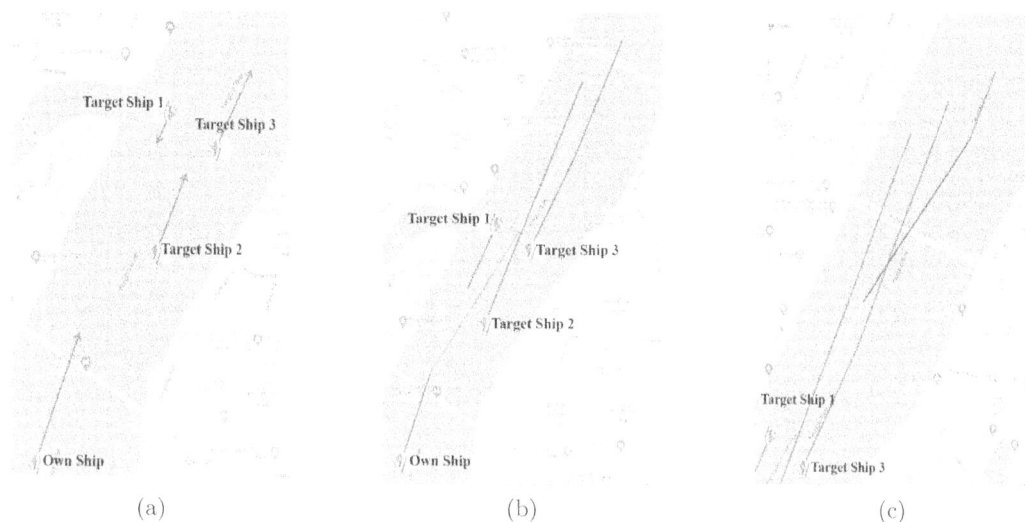

Figure 3.12: Obstacle avoidance trajectory of the own ship simulated according to AIS data in Yangtze River, Wuhan, China. (a) the initial stage of the ships; (b) the two maneuvers to starboard; (c) the last maneuver in the overtaking situation.

with the own ship. This results in a complex multi-ship encountering scenario. Then, according to the proposed algorithm, the own ship calculates the angle of turning with target ship 2. The calculated angle is less than the angle of avoiding target ship 1, and the speed direction remains unchanged, as shown in the red track in the figure. After completely avoiding the target vessel 1, the own ship sails at the speed of avoiding target ship 2, as shown in the yellow track.

In the process of avoiding target ship 2, the own ship detects target ship 3, and is also in a overtaking situation with target ship 3. Then, the own ship selects a larger angle to avoid collision. The purple track represents the collision avoidance route between the own ship and target ship 3, and returns to the initial velocity when the collision avoidance process is completed. Fig. 3.12(c) shows that the own ship avoids all the target ships successfully.

3.8 CONCLUSIONS

In this chapter, the autonomous guidance module of USVs is discussed. Typical guidance methods and their advantages/disadvantages are reviewed. With this overview on guidance techniques in mind and as an illustration, a regulation aware dynamic path planning method is proposed based on uncertain velocity obstacles. The proposed method solves the priority problem in the decision-making process of multi-vessel encountering scenarios. The collision risk of ships is calculated in real time to activate the collision avoidance maneuver. In addition, considering the encountering scenarios of crossing, overtaking and head-on, the idea of robust tube is used to deal with the uncertainties in the target ship information. In the existence of uncertainties, the compliance with COLREGs is also achieved in various encountering scenarios with multiple ships. Simulation experiments and verification with AIS data show that the proposed method can make safe decisions and avoid collisions for intelligent ships in complex marine scenarios.

As stated in the problem formulation, the proposed uncertain VO algorithm needs to know the position and speed of the target ship in advance. Therefore, an effective perception system is necessarily built for the intelligent ship as the one elaborated in Chapter 2. The perception system might be consisted of various onboard sensors, such as lidar and camera, and the fusion of these sensor data.

Bibliography

[1] N. A. Shneydor, *Missile Guidance and Pursuit: Kinematics, Dynamics and Control.* Horwood Publishing Ltd, 1998.

[2] R. Polvara, S. Sharma, J. Wan, A. Manning, and R. Sutton, "Obstacle avoidance approaches for autonomous navigation of unmanned surface vehicles," *The Journal of Navigation*, vol. 71, no. 1, pp. 241–256, 2018.

[3] K. Chu, M. Lee, and M. Sunwoo, "Local path planning for off-road autonomous driving with avoidance of static obstacles," *IEEE Transactions on Intelligent Transportation Systems*, vol. 13, no. 4, pp. 1599–1616, 2012.

[4] M. Zyczkowski and R. Szlapczynski, "Collision risk-informed weather routing for sailboats," *Reliability Engineering & System Safety*, vol. 232, p. 109015, 2023.

[5] B. C. Shah and S. K. Gupta, "Long-distance path planning for unmanned surface vehicles in complex marine environment," *IEEE Journal of Oceanic Engineering*, vol. 45, no. 3, pp. 813–830, 2019.

[6] R. Dubey and S. J. Louis, "VORRT-COLREGs: A hybrid velocity obstacles and RRT based COLREGs-compliant path planner for autonomous surface vessels," in *Proceedings of OCEANS 2021*, San Diego, Porto, USA, 2021, pp. 1–8.

[7] C. Shi, M. Zhang, and J. Peng, "Harmonic potential field method for autonomous ship navigation," in *Proceedings of the 2007 7th International Conference on ITS Telecommunications*, Sophia Antipolis, France, 2007, pp. 1–6.

[8] E. Alagić, J. Velagić, and A. Osmanović, "Design of mobile robot motion framework based on modified vector field histogram," in *Proceedings of International Symposium ELMAR*, Zadar, Croatia, 2019, pp. 135–138.

[9] Y. Li and Q. Zhu, "Local path planning based on improved dynamic window approach," in *Proceedings of the 40th Chinese Control Conference*, Shanghai, China, 2021, pp. 4291–4295.

[10] H. Zheng, J. Zhu, C. Liu, H. Dai, and Y. Huang, "Regulation aware dynamic path planning for intelligent ships with uncertain velocity obstacles," *Ocean Engineering*, vol. 278, p. 114401, 2023.

[11] N. Wang and H. Xu, "Dynamics-constrained global-local hybrid path planning of an autonomous surface vehicle," *IEEE Transactions on Vehicular Technology*, vol. 69, no. 7, pp. 6928–6942, 2020.

[12] A. Tsolakis, R. R. Negenborn, V. Reppa, and L. Ferranti, "Model predictive trajectory optimization and control for autonomous surface vessels considering traffic rules," *IEEE Transactions on Intelligent Transportation Systems*, 2024.

[13] Z. He, C. Liu, X. Chu, R. R. Negenborn, and Q. Wu, "Dynamic anti-collision A-star algorithm for multi-ship encounter situations," *Applied Ocean Research*, vol. 118, p. 102995, 2022.

[14] X. Wu, Y. Wang, S. Chai, and R. Chai, "Goal-biased rapidly-exploring random trees for efficient marine path planning," in *Proceedings of China Automation Congress*, Chongqing, China, 2023, pp. 4674–4679.

[15] X. Tong, S. Yu, G. Liu, X. Niu, C. Xia, J. Chen, Z. Yang, and Y. Sun, "A hybrid formation path planning based on a* and multi-target improved artificial potential field algorithm in the 2D random environments," *Advanced Engineering Informatics*, vol. 54, p. 101755, 2022.

[16] Z. He, X. Chu, C. Liu, and W. Wu, "A novel model predictive artificial potential field based ship motion planning method considering COLREGs for complex encounter scenarios," *ISA Transactions*, vol. 134, pp. 58–73, 2023.

[17] D. Ma, S. Hao, W. Ma, H. Zheng, and X. Xu, "An optimal control-based path planning method for unmanned surface vehicles in complex environments," *Ocean Engineering*, vol. 245, p. 110532, 2022.

[18] S. L. B. S. W. X. Guoshun Liu, Huarong Zheng, "Hierarchical dynamic trajectory planning for autonomous underwater vehicles: Algorithms and experiments," *Ocean Engineering*, vol. 307, p. 118221, 2024.

[19] X. Guo, M. Ji, Z. Zhao, D. Wen, and W. Zhang, "Global path planning and multi-objective path control for unmanned surface vehicle based on modified particle swarm optimization (PSO) algorithm," *Ocean Engineering*, vol. 216, p. 107693, 2020.

[20] L. Li, D. Wu, Y. Huang, and Z.-M. Yuan, "A path planning strategy unified with a colregs collision avoidance function based on deep reinforcement learning and artificial potential field," *Applied Ocean Research*, vol. 113, p. 102759, 2021.

[21] N. Wang, Y. Zhang, C. K. Ahn, and Q. Xu, "Autonomous pilot of unmanned surface vehicles: Bridging path planning and tracking," *IEEE Transactions on Vehicular Technology*, vol. 71, no. 3, pp. 2358–2374, 2022.

[22] S. Mahmoud Zadeh, A. Abbasi, A. Yazdani, H. Wang, and Y. Liu, "Uninterrupted path planning system for multi-USV sampling mission in a cluttered ocean environment," *Ocean Engineering*, vol. 254, p. 111328, 2022.

[23] R. Nian, Z. Shen, and W. Ding, "Research on global path planning of unmanned sailboat based on improved ant colony optimization," in *Proceedings of the 6th International Conference on Automation, Control and Robotics Engineering*, Dalian, China, 2021, pp. 428–432.

[24] Y. Yu, H. Zheng, and W. Xu, "Learning and sampling-based informative path planning for AUVs in ocean current fields," *IEEE Transactions on Systems, Man, and Cybernetics: Systems*, vol. 55, no. 1, pp. 51–62, 2025.

[25] C. Chen, X.-Q. Chen, F. Ma, X.-J. Zeng, and J. Wang, "A knowledge-free path planning approach for smart ships based on reinforcement learning," *Ocean Engineering*, vol. 189, p. 106299, 2019.

[26] J. Woo and N. Kim, "Collision avoidance for an unmanned surface vehicle using deep reinforcement learning," *Ocean Engineering*, vol. 199, p. 107001, 2020.

[27] L. Zhao, T. Szakas, M. Churley, and D. Lucas, "Multi-class multi-residue analysis of pesticides in edible oils by gas chromatography-tandem mass spectrometry using liquid-liquid extraction and enhanced matrix removal lipid cartridge cleanup," *Journal of Chromatography A*, vol. 1584, pp. 1–12, 2019.

[28] J. Ma, J. Jin, D. Liu, L. Liu, D. Wang, and M. Chen, "An improved adaptive dynamic window algorithm for target tracking of unmanned surface vehicles," in *Proceedings of International Conference on Signal Processing, Communications and Computing*, Xi'an, China, 2021, pp. 1–5.

[29] T. I. Fossen, M. Breivik, and R. Skjetne, "Line-of-sight path following of underactuated marine craft," in *Proceedings of the 6th IFAC on Manoeuvring and Control of Marine Craft*, Girona, Spain, 2003, pp. 244–249.

[30] Y. Jiang, Z. Peng, D. Wang, and C. P. Chen, "Line-of-sight target enclosing of an underactuated autonomous surface vehicle with experiment results," *IEEE Transactions on Industrial Informatics*, vol. 16, no. 2, pp. 832–841, 2019.

[31] Z. Tian, H. Zheng, and W. Xu, "Path following of autonomous surface vehicles with line-of-sight and nonlinear model predictive control," in *Proceedings of the 6th International Conference on Transportation Information and Safety*, Wuhan, China, 2021, pp. 1269–1274.

[32] Y. Zhong, J. Liu, J. Huang, L. Bao, and Z. Wang, "Unmanned vehicle target tracking method based on dubins and constant bearing guidance," in *Proceedings of International Conference on Mechatronics and Automation*, Harbin, China, 2023, pp. 826–832.

[33] L. Liu, D. Wang, and Z. Peng, "ESO-based line-of-sight guidance law for path following of underactuated marine surface vehicles with exact sideslip compensation," *IEEE Journal of Oceanic Engineering*, vol. 42, no. 2, pp. 477–487, 2017.

[34] R. Rout, R. Cui, and W. Yan, "Sideslip-compensated guidance-based adaptive neural control of marine surface vessels," *IEEE Transactions on Cybernetics*, vol. 52, no. 5, pp. 2860–2871, 2022.

[35] Z. Zheng, "Moving path following control for a surface vessel with error constraint," *Automatica*, vol. 118, p. 109040, 2020.

[36] L. Wan, Y. Su, H. Zhang, B. Shi, and M. S. AbouOmar, "An improved integral light-of-sight guidance law for path following of unmanned surface vehicles," *Ocean Engineering*, vol. 205, p. 107302, 2020.

[37] F. Liu, Y. Shen, B. He, D. Wang, J. Wan, Q. Sha, and P. Qin, "Drift angle compensation-based adaptive line-of-sight path following for autonomous underwater vehicle," *Applied Ocean Research*, vol. 93, p. 101943, 2019.

[38] L. Liu, D. Wang, Z. Peng, C. L. P. Chen, and T. Li, "Bounded neural network control for target tracking of underactuated autonomous surface vehicles in the presence of uncertain target dynamics," *IEEE Transactions on Neural Networks and Learning Systems*, vol. 30, no. 4, pp. 1241–1249, 2019.

[39] P. N. F. b. Mohd Shamsuddin and M. A. b. Mansor, "Motion control algorithm for path following and trajectory tracking for unmanned surface vehicle: A review paper," in *Proceedings of the 3rd International Conference on Control, Robotics and Cybernetics*, Penang, Malaysia, 2018, pp. 73–77.

[40] N. Gu, D. Wang, Z. Peng, J. Wang, and Q.-L. Han, "Advances in line-of-sight guidance for path following of autonomous marine vehicles: An overview," *IEEE Transactions on Systems, Man, and Cybernetics: Systems*, vol. 53, no. 1, pp. 12–28, 2023.

[41] Z. Peng, J. Wang, and D. Wang, "Distributed maneuvering of autonomous surface vehicles based on neurodynamic optimization and fuzzy approximation," *IEEE Transactions on Control Systems Technology*, vol. 26, no. 3, pp. 1083–1090, 2018.

[42] H. Zheng, R. R. Negenborn, and G. Lodewijks, "Fast ADMM for distributed model predictive control of cooperative waterborne AGVs," *IEEE Transactions on Control Systems Technology*, vol. 25, no. 4, pp. 1406–1413, Jul. 2017.

[43] U.S. Dept. Homeland Security/U.S. Coast Guard, "Navigation rules," Paradise Cay Publications, 2010.

[44] R. Polvara, S. Sharma, J. Wan, A. Manning, and R. Sutton, "Obstacle avoidance approaches for autonomous navigation of unmanned surface vehicles," *The Journal of Navigation*, vol. 71, no. 1, pp. 241–256, Jan. 2018.

[45] K. Chu, M. Lee, and M. Sunwoo, "Local path planning for off-road autonomous driving with avoidance of static obstacles," *IEEE Transactions on Intelligent Transportation Systems*, vol. 13, no. 4, pp. 1599–1616, Dec. 2012.

[46] B. C. Shah and S. K. Gupta, "Long-distance path planning for unmanned surface vehicles in complex marine environment," *IEEE Journal of Oceanic Engineering*, vol. 45, no. 3, pp. 813–830, Jul. 2020.

[47] H. Wang, W. Mao, and L. Eriksson, "A Three-Dimensional Dijkstra's algorithm for multi-objective ship voyage optimization," *Ocean Engineering*, 2019, 186, 106131.

[48] A. Lazarowska, "Ship's trajectory planning for collision avoidance at sea based on ant colony optimisation," *The Journal of Navigation*, vol. 68, no. 2, pp. 291–307, 2015.

[49] J. Xin, C. Meng, A. D'Ariano, F. Schulte, J. Peng*, and R.R. Negenborn, "Energy-Efficient Routing of a Multi-Robot Station: A Flexible Time-Space Network Approach," *IEEE Transactions on Automation Science and Engineering*, DOI:10.1109/TASE.2022.3192914

[50] C. Shi, M. Zhang, and J. Peng, "Harmonic potential field method for autonomous ship navigation," *in Proceedings of the 2007 IEEE International Conference on ITS Telecommunications*, Sophia Antipolis, France, 2007, pp. 1–6.

[51] Z. Zuo, X. Yang, Z. Li, Y. Wang, Q. Han, L. Wang, and X. Luo, "MPC-based cooperative control strategy of path planning and trajectory tracking for intelligent vehicles," *IEEE Transactions on Intelligent Vehicles*, vol. 6, no. 3, pp. 513–522, Sep. 2021.

[52] J. Borenstein and Y. Koren, "The vector field histogram—Fast obstacle avoidance for mobile robots," *IEEE Transactions on Robotics and Automation*, vol. 7, no. 3, pp. 278–288, Jun. 1991.

[53] S. Han, L. Wang, Y. Wang, and H. He, "A dynamically hybrid path planning for unmanned surface vehicles based on non-uniform Theta* and improved dynamic windows approach," *Ocean Engineering*, 2022, vol. 257, no. 111655.

[54] P. Fiorini and Z. Shiller, "Motion planning in dynamic environments using velocity obstacles," *The International Journal of Robotics Research*, vol. 17, no. 7, pp.760–772, 1998.

[55] J. van den Berg, M. Lin, and D. Manocha, "Reciprocal velocity obstacles for real-time multi-agent navigation," *in Proceedings of the 2008 IEEE international conference on robotics and automation*, Pasadena, CA, May 2008, pp. 1928–1935.

[56] Y. Huang, L. Chen, and P. van Gelder, "Generalized velocity obstacle algorithm for preventing ship collisions at sea," *Ocean Engineering*, vol. 173, pp. 142–156, Feb. 2019.

[57] J. Snape, J. Van den Berg, S. J. Guy, and D. Manocha, "The hybrid reciprocal velocity obstacle," *IEEE Transactions on Robotics*, vol. 27, no. 4, pp. 696–706, Aug. 2011.

[58] Y. Kuwata, M. T. Wolf, D. Zarzhitsky and T. L. Huntsberger, "Safe maritime autonomous navigation with COLREGs, using velocity obstacles," *IEEE Journal of Oceanic Engineering*, vol. 39, no. 1, pp. 110-119, 2014.

[59] S.-M. Lee, K.-Y. Kwon, and J. Joh, "A fuzzy logic for autonomous navigation of marine vehicles satisfying COLREG guidelines," *International Journal of Control, Automation, and Systems*, vol. 2, no. 2, pp. 171–181, 2004.

[60] M. Benjamin, J. Curcio, and P. Newman, "Navigation of unmanned marine vehicles in accordance with the rules of the road," *in Proceedings of the 2006 IEEE International Conference on Robotics and Automation*, Orlando, FL, USA, 2006, pp. 3581–3587.

[61] Y. Wang, X. Yu, X. Liang, and B. Li, "A COLREGs-based obstacle avoidance approach for unmanned surface vehicles," *Ocean Engineering*, vol. 169, pp. 110–124, Dec. 2018.

[62] H. Lyu and Y. Yin, "COLREGS-constrained real-time path planning for autonomous ships using modified artificial potential fields," *The Journal of Navigation*, vol. 72, no. 3, pp. 588–608, May 2019.

[63] J. Zhang, D. Zhang, X. Yan, S. Haugen, and C. G. Soares, "A distributed anti-collision decision support formulation in multi-ship encounter situations under COLREGs," *Ocean Engineering*, vol. 105, pp. 336–348, Sep. 2015.

[64] L. Hu, W. Naeem, E. Rajabally, G. Watson, T. Mills, Z. Bhuiyan, C. Raeburn, I. Salter, and C. Pekcan, "A multiobjective optimization approach for COLREGs-compliant path planning of autonomous surface vehicles verified on networked bridge simulators," *IEEE Transactions on Intelligent Transportation Systems*, vol. 21, no. 3, pp. 1167–1179, Mar. 2020.

[65] T. A. Johansen, T. Perez, and A. Cristofaro, "Ship collision avoidance and COLREGS compliance using simulation-based control behavior selection with predictive hazard assessment," *IEEE Transactions on Intelligent Transportation Systems*, vol. 17, no. 12, pp. 3407–3422, Dec. 2016.

[66] L. Zhao and X. Fu, "A Method for Correcting the Closest Point of Approach Index During Vessel Encounters Based on Dimension Data From AIS," *IEEE Transactions on Intelligent Transportation Systems*, doi: 10.1109/TITS.2021.3127223.

[67] A. K. Debnath and H. C. Chin, "Navigational traffic conflict technique: A proactive approach to quantitative measurement of collision risks in port waters," *The Journal of Navigation*, vol. 63, no. 01, pp. 137–152, 2010.

[68] L. Zhang and Q. Meng, "Probabilistic ship domain with applications to ship collision risk assessment," *Ocean Engineering*, vol. 186, Aug. 2019, Art. no. 106130.

[69] Z. Pietrzykowski and M. Wielgosz, "Effective ship domain-impact of ship size and speed," *Ocean Engineering*, vol. 219, p. 108423, 2021.

[70] J. Park and J. Kim, "Predictive evaluation of ship collision risk using the concept of probability flow," *IEEE Journal of Occanic Engineering*, vol. 42, no. 4, pp. 836–845, Oct. 2017.

[71] A. Majumdar and R. Tedrake, "Funnel libraries for real-time robust feedback motion planning," *The International Journal of Robotics Research*, vol. 36, no. 8, pp. 947–982, 2017.

[72] L. Gang, Y. Wang, Y. Sun, L. Zhou, and M. Zhang, "Estimation of vessel collision risk index based on support vector machine," *Advances in Mechanical Engineering*, vol. 8, no. 11, pp. 1–10, 2016.

[73] B. Wu, T. Cheng, T. L. Yip, and Y. Wang, "Fuzzy logic based dynamic decision-making system for intelligent navigation strategy within inland traffic separation schemes," *Ocean Engineering*, vol. 197, Feb. 2020, Art. no. 106909.

[74] R. Song, Y. Liu, and R. Bucknall, "Smoothed A* algorithm for practical unmanned surface vehicle path planning," *Applied Ocean Research*, 83:9–20, 2019.

[75] Y. Cheng and W. Zhang, "Concise deep reinforcement learning obstacle avoidance for underactuated unmanned marine vessels," *Neurocomputing*, vol. 272, pp. 63–73, Jan. 2018.

[76] L. Zhao and M.-I. Roh, "COLREGs-compliant multiship collision avoidance based on deep reinforcement learning," *Ocean Engineering*, vol. 191, Nov. 2019, Art. no. 106436.

[77] S. Han, L. Wang, and Y. Wang, " A potential field-based trajectory planning and tracking approach for automatic berthing and COLREGs-compliant collision avoidance," *Ocean Engineering*, vol. 266, p. 112877, 2022.

[78] P. V. Davis, M. J. Dove, and C. T. Stockel, "A computer simulation of marine traffic using domains and arenas," *The Journal of Navigation*, vol. 33, pp. 215–222, 1980.

[79] C. Fulgenzi, A. Spalanzani, and C. Laugier, "Dynamic obstacle avoidance in uncertain environment combining PVOs and occupancy grid," *in Proceedings of the 2007 IEEE International Conference on Robotics and Automation (ICRA)*, Roma, Italy, 2007, pp. 1610–1616.

[80] N. Wang, S. Lv, M. J. Er, and W. H. Chen, "Fast and accurate trajectory tracking control of an autonomous surface vehicle with unmodeled dynamics and disturbances," *IEEE Transactions on Intelligent Vehicles*, vol. 1, no. 3, pp. 230–243, Sep. 2016.

CHAPTER 4

USV Control Methods and Robust Control Handling Maritime Disturbances

4.1 INTRODUCTION

FOR USVs, the control module is a closed-loop feedback subsystem that involves sophisticated algorithms to achieve the desired USV behaviors. The control system uses the filtered system outputs from the navigation module, compares the outputs with the references from the guidance module, computes corrective inputs probably based on the USV dynamics model to a certain degree, and applies the inputs to the USV system.

In general, the USV control tasks can be categorized into three types, i.e., setpoint regulation, path following and trajectory tracking. Various control algorithms have been proposed to achieve these control tasks. The control techniques can be roughly categorized into model-based control, model free control and hybrid control. We note that, compared to other control systems, USVs are featured with special issues that have in turn progressed the field of control. One of the major issues is that USVs face complex marine environment disturbances, which will also be considered in this chapter.

4.2 USV CONTROL PROBLEMS

According to the different purposes, the commonly seen USV control problems include the speed control [1][2], heading control [3], and the more general motion control [4, 5, 6]. The lower level control allocation and actuator control problems [7] are also important for USVs. Most of the existing research focuses on the relatively more sophisticated motion control problems of marine surface vehicles. Particularly, three categories of motion control problems with different purposes can be recognized [8].

- **Setpoint regulation**. In this case, the references to the USV control system are constant, and the corresponding controller is also called a regulator.

DOI: 10.1201/9781003660590-4

Examples are constant speed regulation [1] and dynamic positioning [9]. Due to the ubiquitously existing environmental disturbances, the robustness against uncertainties [9, 10, 11] is especially important in the setpoint regulation problems of USVs.

- **Path following**. The reference is a geometric path independent of time. The reference path can be straight lines [12, 4, 13] or curves [14][15] without temporal constraints. The common approach is to parameterize the geometric path with virtual dynamics [4], of which the guidance problem can be combined with the tracking control of USVs. Commonly seen guidance approaches are LOS guidance [16, 17, 12, 13] and pure pursuit [18].

- **Trajectory tracking**. Explicit time differentiable references (e.g., positions, velocities) must be given in the trajectory tracking problems [19, 20]. These time dependent references can be generated via path parameterization [4] as mentioned before or by reference generators [21]. The control goal is then to drive the system to specified states at specified time with minimal tracking errors.

Overall, there are mainly three types of USV control design ideas, i.e., model-based, model free and hybrid control, to achieve the aforementioned control tasks. If preliminary understanding of the USV system, e.g., the USV mathematical models, are available, model-based control laws can be derived. Otherwise, model free control actions can be obtained via, e.g., experiences, trials-and-errors or pattern recognition from sufficient system data. Different types of control techniques can also be combined as hybrid USV control methods. In this chapter, we will first focus on model-based USV control methods. The model free and hybrid control methods will be discussed in the next chapter.

4.3 MODEL-BASED USV CONTROL METHODS

Model-based USV control methods generally employ the three degree-of-freedom (DOF) mass-damping-Coriolis surface vessel dynamics model [8, 5]. The three DOF model considers detailed hydrostatics, hydrodynamics, and kinematics with environmental influences in the inertial and body-fixed coordinate systems, and involves highly nonlinear terms. The main maneuvering characteristics of USVs can be captured by this model.

However, the hydrodynamic parameters could be time-varying due to the varying system and environmental states. Sophisticated system identification methods [22, 23, 24, 25, 26] are thus usually necessary to establish the mathematical model of a specified USV platform. To save the efforts in the system identification process, the open source platform models such as Cybership II [27] can also be used for simulation purposes [5]. Special designs are usually necessary to take care of the system model deficiencies such as underactuation [28, 29, 30, 31, 32] or model mismatches [26, 33, 20]. With the knowledge on the system dynamics, various model-based USV control methods have been proposed.

The first category of model-based control methods mainly focus on the closed-loop stability properties. These methods mostly analyze the nonlinear USV dynamics characteristics and design the so-called Lyapunov function [34, 35] directly or indirectly. The closed-loop system stability is guaranteed, which usually means that the USV system would be stabilized at a position (i.e., the setpoint regulation task) or the USV tracking error system would be stabilized at the origin (i.e., the path following or trajectory tracking tasks). The sliding mode control and its variants, due to the robustness against small magnitude of disturbances, have seen widely applications in USV control problems [36, 37, 38, 39, 24]. Sliding mode control laws are discontinuous and could have the "chattering" phenomenon. Another popular nonlinear control method for USVs is the backstepping design [40, 25]. The autonomous ship group at Norwegian University of Science and Technology (NTNU) also utilizes the backstepping control extensively [12, 33, 41]. The idea is to divide the complex system into multiple sub-systems for which virtual control laws are designed and combined to obtain the system control law. The maneuvering problem of USVs is proposed in [33] where a geometric task and a dynamic task are involved. The geometric task guarantees path convergence and the dynamic task tracks an assigned speed along the path. For these nonlinear control methods, analytical control laws are usually available so the computational burden is acceptable for real-time applications. However, system constraints are generally considered with the saturation design, and the cost performance cannot be systematically considered as well.

The second category of model-based methods focuses explicitly on the optimal performance, e.g., Linear-Quadratic-Gaussian (LQG) controller [42] and model predictive control (MPC) [4]. Optimal control [43] differs with other control techniques in that it can formulate a particular *objective function*, and thus achieves desired behaviors with an optimal cost. Performance indices such as tracking errors, energy or time consumption can be optimized. Notably, MPC, as a type of optimal control based control methods, can take into account the constraints and cost indices explicitly, and thus have been widely applied for USV control problems [44, 45, 4, 46]. Recently, MPC has been applied to vessel path following [13, 47], trajectory tracking [48] and heading control [49]. However, optimal control relies on solving mathematical optimization problems which could be hard when nonlinear USV system dynamics and constraints are involved.

4.4 ROBUST MODEL PREDICTIVE CONTROL METHODS

System robustness in face of uncertainties is desirable in order to maintain certain system performance and constraint satisfactions. In addition to systematically handling system constraints and optimizing performance, MPC [50], under certain conditions, is inherently robust due to repetitively online open-loop optimizations with newly measured system states. But this inherent robustness is only against sufficiently small uncertainties [51]. A more reliable approach is to have designed robustness by taking all possible realizations of uncertainties as well as the propagation of the effects of uncertainties over the time horizon into account. However, considering uncertainties in open-loop predictions can lead to a large bunch of trajectories, which further lead to conservativeness and even infeasibility. Incorporating feedbacks into predictions and optimizing over control policies rather than control sequences in online optimizations [52] have, therefore, become standard practice for controlling uncertain systems using MPC.

However, optimizing over arbitral control policies is practically intractable. Approximations such as affine state or disturbance feedback parameterizations of the control policy have been made [53, 54, 55]. These approaches are computationally tractable but still heavy since decision variables include elements of feedback matrices. Tube-based robust MPC (RMPC) which has the same order of complexity as that of conventional MPC parameterizes the control policy with an open-loop control sequence and a local feedback [56]. The tube containing all possible system trajectories uses a nominal trajectory as its center and uncertainties that are steered to the center by the local feedback as cross sections. Rigorous RMPC synthesis has been derived using the above different parametrization schemes for a class of constrained linear discrete-time invariant system with bounded disturbances [53, 56, 54, 55].

However, uncertain environmental forces and moments in vessel motion control problems often bear stochastic characteristics [1]. Since constrained systems cannot be robust for all uncertainty realizations, in the case of unbounded stochastic uncertainties, worst-case hard constraints as imposed in RMPC are usually softened either by average (expected)-case constraints [57] or by chance constraints [58]. Stochastic uncertainties are either made bounded [59] or truncated [60] for use in RMPC, but still, bounds or truncations are fixed in online optimization. If the bounds are set too large, system performance or even solvability is not guaranteed; if the bounds are set too small, a certain level of robustness or safety cannot be achieved.

4.5 TUBE-BASED ROBUST MODEL PREDICTIVE CONTROL

In this section, we propose an integrated bounding of stochastic uncertainties in RMPC for USVs. In particular, we model uncertainties with known probabilistic distributions by introducing binary variables. Binary variables representing probabilistic information arise in both constraints and cost function of a tube-based RMPC problem. Since both constraint tightenings and cost functions of online optimizations are parameterized explicitly by uncertainty distributions, system capability, uncertainty properties, and performance criteria are systematically integrated providing a trade-off among system performance, problem solvability, and robustness attributing to a cost-effective approach.

We first apply successive linearizations to obtain approximated linear time-varying models and constraints. Then, a tube-based RMPC controller is proposed with explicit use of probabilistic distributions of the stochastic disturbances.

4.5.1 Successive linearizations

We inherit the successive linearization approach in [4]. Interested readers are refereed to [4] for more details. For notational consistency and simplicity, USVs dynamics are generalized and discretized with the zero-order-hold assumption as:

$$\boldsymbol{x}(k+1) = f_{\mathrm{d}}(\boldsymbol{x}(k), \boldsymbol{u}(k), \boldsymbol{b}(k)) \tag{4.1}$$

with system states $\boldsymbol{x}(k) \in \mathbb{R}^6$, control inputs $\boldsymbol{u}(k) \in \mathbb{R}^3$ and disturbance inputs $\boldsymbol{b}(k) \in \mathbb{R}^3$. Discrete time k denotes real time kT_{s} with sampling time T_{s}. Successive linearizations of (4.1) are primarily implemented in three steps at each discrete time:

1. Shift the previous optimal control input sequence to obtain a seed control input trajectory $\boldsymbol{u}^0(i|k)$ for $i = 0, 1, ..., N_{\mathrm{p}} - 1$[1] where N_{p} is the prediction horizon;

2. Apply $\boldsymbol{u}^0(i|k)$ to (4.1) and obtain a seed state trajectory $\boldsymbol{x}^0(i|k)$. However, since decisions are made after knowing the current state but before knowing the disturbance inputs, initial states are set as $\boldsymbol{x}^0(0|k) = \boldsymbol{x}(k)$ while $\boldsymbol{b}(i|k)$ remain unknown. The predicted mean values are utilized to define the seed disturbance input trajectory as $\boldsymbol{b}^0(i|k) = \bar{\boldsymbol{b}}(i|k)$. Therefore, $\boldsymbol{x}^0(i+1|k) = f_{\mathrm{d}}\left(\boldsymbol{x}^0(i|k), \boldsymbol{u}^0(i|k), \boldsymbol{b}^0(i|k)\right)$;

3. Approximate nonlinear dynamics (4.1) by linearizing about seed trajectories $\left(\boldsymbol{x}^0(i|k), \boldsymbol{u}^0(i|k), \boldsymbol{b}^0(i|k)\right)$ as:

$$\boldsymbol{x}(i{+}1|k) = \boldsymbol{x}^0(i{+}1|k) + \boldsymbol{A}(i|k)\Delta\boldsymbol{x}(i|k) + \boldsymbol{B}(i|k)\Delta\boldsymbol{u}(i|k) + \boldsymbol{E}(i|k)\Delta\boldsymbol{b}(i|k), \tag{4.2}$$

with $\Delta\boldsymbol{x}(i|k), \Delta\boldsymbol{u}(i|k), \Delta\boldsymbol{b}(i|k)$ being small perturbations around $\boldsymbol{x}^0(i|k)$, $\boldsymbol{u}^0(i|k)$ and $\boldsymbol{b}^0(i|k)$, respectively; and $\boldsymbol{A}(i|k)$, $\boldsymbol{B}(i|k)$ and $\boldsymbol{E}(i|k)$ are Jacobian matrices with respect to $\boldsymbol{x}$, $\boldsymbol{u}$ and $\boldsymbol{b}$, respectively, of function f_{d} evaluated at $\left(\boldsymbol{x}^0(i|k), \boldsymbol{u}^0(i|k), \boldsymbol{b}^0(i|k)\right)$. In a similar way, approximate non-convex obstacle avoidance constraint as:

$$d(i|k) = d^0(i|k) + \boldsymbol{C}(i|k)\Delta x(i|k) + \boldsymbol{D}(i|k)\Delta y(i|k), \tag{4.3}$$

for $i = 1, 2, ..., N_{\mathrm{p}}$ with $\boldsymbol{C}(i|k)$ and $\boldsymbol{D}(i|k)$ being Jacobian matrices of function d with respect to x and y, respectively, evaluated at $\left(x^0(i|k), y^0(i|k)\right)$.

Two of the prerequisites for a tube-based formulation [56] are: *1*) linearity of the system so that the partition is possible; and *2*) compactness of the uncertainties so that the cross sections are bounded. The first prerequisite is satisfied by using the approximated linear model (4.2) and constraint (4.3) instead of (4.1). The second is satisfied by assuming that robustness is only against a certain probability p of the infinite support stochastic uncertainties. In such cases, a controller can be designed to be robust to 95% of disturbances b, which means $b \in \left[-\sqrt{2}\mathrm{erf}^{-1}(95\%), \sqrt{2}\mathrm{erf}^{-1}(95\%)\right] = [-2, 2]$ if the disturbance is following a standard normal distribution [61].

---

[1]The prediction step i for control/disturbance inputs over $0, 1, ..., N_{\mathrm{p}} - 1$ and for state over $0, 1, ..., N_{\mathrm{p}}$ is omitted for simplicity in the remainder of this chapter if contextually clear.

4.5.2 Tube-based RMPC

We partition uncertain system states as:

$$\boldsymbol{x}(i|k) = \bar{\boldsymbol{x}}(i|k) + \Delta\tilde{\boldsymbol{x}}(i|k) = \boldsymbol{x}^0(i|k) + \Delta\bar{\boldsymbol{x}}(i|k) + \Delta\tilde{\boldsymbol{x}}(i|k), \tag{4.4}$$

where $\bar{\boldsymbol{x}}$ denotes the nominal state; $\Delta\bar{\boldsymbol{x}}$ is the nominal perturbation state and $\Delta\tilde{\boldsymbol{x}}$ the deviation of the actual perturbation state $\Delta\boldsymbol{x}$ from $\Delta\bar{\boldsymbol{x}}$. Likewise, for control input,

$$\boldsymbol{u}(i|k) = \bar{\boldsymbol{u}}(i|k) + \Delta\tilde{\boldsymbol{u}}(i|k) = \boldsymbol{u}^0(i|k) + \Delta\bar{\boldsymbol{u}}(i|k) + \Delta\tilde{\boldsymbol{u}}(i|k).$$

Then, uncertain perturbation dynamics evolve as:

$$\Delta\tilde{\boldsymbol{x}}(i+1|k) = \boldsymbol{A}(i|k)\Delta\tilde{\boldsymbol{x}}(i|k) + \boldsymbol{B}(i|k)\Delta\tilde{\boldsymbol{u}}(i|k) + \boldsymbol{E}(i|k)\Delta\boldsymbol{b}(i|k), \tag{4.5}$$

with $\Delta\tilde{\boldsymbol{x}}(0|k) = \boldsymbol{0}$ since $\bar{\boldsymbol{x}}(0|k) = \boldsymbol{x}^0(0|k) = \boldsymbol{x}(k)$; $\Delta\boldsymbol{b} \in \mathbb{W}$ where $\mathbb{W}$ is a compact set with origin in its interior. We employ a linear parametrization for the control law as:

$$\Delta\tilde{\boldsymbol{u}}(i|k) = \boldsymbol{K}(i|k)\Delta\tilde{\boldsymbol{x}}(i|k), \tag{4.6}$$

where $\boldsymbol{K}(i|k)$ is calculated online as an optimal LQR feedback gain for unconstrained time-varying linear dynamics (4.5) as in [62]. Then, (4.5) in closed-loop is:

$$\Delta\tilde{\boldsymbol{x}}(i+1|k) = \boldsymbol{A_K}(i|k)\Delta\tilde{\boldsymbol{x}}(i|k) + \boldsymbol{E}(i|k)\Delta\boldsymbol{b}(i|k), \tag{4.7}$$

with $\boldsymbol{A_K}(j|k) = \boldsymbol{A}(i|k) + \boldsymbol{B}(i|k)\boldsymbol{K}(i|k)$. Denote uncertain perturbation sets as $\tilde{\mathbb{X}}_\Delta$, i.e., $\Delta\tilde{\boldsymbol{x}}(i|k) \in \tilde{\mathbb{X}}_\Delta(i|k)$, we further have:

$$\tilde{\mathbb{X}}_\Delta(i+1|k) := \boldsymbol{A_K}(i|k)\tilde{\mathbb{X}}_\Delta(i|k) \oplus \boldsymbol{E}(i|k)\mathbb{W}, \tag{4.8}$$

with $\tilde{\mathbb{X}}_\Delta(0|k) = \{\boldsymbol{0}\}$. The sets $\{\tilde{\mathbb{X}}_\Delta(i|k)\}$ are expected to be smaller than those calculated from unstable pairs of $(\boldsymbol{A}(i|k), \boldsymbol{B}(i|k))$ since $\boldsymbol{A_K}(i|k)$ is now stable by design. The operator $\oplus$ denotes the Minkowski set sum defined as $A \oplus B := \{a+b | a \in A, b \in B\}$. Therefore, the tube $\{\mathbb{X}(i|k)\}$ with $\{\bar{\boldsymbol{x}}(i|k)\}$ as the center and $\{\tilde{\mathbb{X}}_\Delta(i|k)\}$ as cross sections is:

$$\mathbb{X}(i|k) := \bar{\boldsymbol{x}}(i|k) \oplus \tilde{\mathbb{X}}_\Delta(i|k). \tag{4.9}$$

Denote $\mathbb{C}_x$ and $\mathbb{C}_u$ as system state and input constraint sets, respectively, imposed by actual system constraints, a nominal optimal predictive control problem for USVs fulfilling the path tracking tasks as in [4] is then readily formulated with tightened constraints as:

$$(\Delta\bar{\boldsymbol{x}}^*(i|k), \Delta\bar{\boldsymbol{u}}^*(i|k)) = \underset{\Delta\bar{\boldsymbol{x}}(i|k),\Delta\bar{\boldsymbol{u}}(i|k)}{\arg\min} \left\{ \bar{J}\left(\boldsymbol{x}(k), \Delta\bar{\boldsymbol{u}}(i|k)\right) \mid \right. \tag{4.10}$$

$$\left. \Delta\bar{\boldsymbol{x}}(i|k) \in \mathbb{C}_x \ominus \tilde{\mathbb{X}}_\Delta(i|k) \ominus \boldsymbol{x}^0(i|k), \Delta\bar{\boldsymbol{u}}(i|k) \in \mathbb{C}_u \ominus \boldsymbol{K}(i|k)\tilde{\mathbb{X}}_\Delta(i|k) \ominus \boldsymbol{u}^0(i|k) \right\},$$

where $\bar{J}\left(\boldsymbol{x}(k), \Delta\bar{\boldsymbol{u}}(i|k)\right)$ is a nominal cost function designed as in [4] to achieve tracking tasks. The operator $\ominus$ denotes Minkowski set difference defined as $A \ominus B := \{a - b | a \in A, b \in B\}$.

4.5.3 Implementations

Solving problem (4.10) requires set computations $\oplus$ and $\ominus$, which are time consuming. Unfortunately, due to time varying nature of the system (4.2) and constraint (4.3), set computations are necessarily conducted online. Therefore, implemented tube-based RMPC treats each constraint separately. Suppose there are $\mathbb{I}_x$ state constraints and for $j = 1, ..., \mathbb{I}_x$:

$$\boldsymbol{C}_j(i|k)\Delta\bar{\boldsymbol{x}}(i|k) + \boldsymbol{C}_j(i|k)\Delta\tilde{\boldsymbol{x}}(i|k) \leqslant d_j(i|k) - d_j^0(i|k), \tag{4.11}$$

then it would be sufficient to tighten the nominal term via an offset defined as the out-bounding of the uncertainty term:

$$
\begin{aligned}
d_x^j(i|k) &:= \max_{\Delta\boldsymbol{b}(i|k)} \left\{ \boldsymbol{C}_j(i|k)\Delta\tilde{\boldsymbol{x}}(i|k) \big| \Delta\tilde{\boldsymbol{x}}(i|k) \in \tilde{\mathbb{X}}_\Delta(i|k) \right\} \\
&:= \max_{\Delta\boldsymbol{b}(i|k)} \left\{ \boldsymbol{C}_j(i|k)\boldsymbol{\Theta}(i|k)\boldsymbol{\Phi}(i|k) \big| \Delta\boldsymbol{b}(i|k) \in \mathbb{W} \right\},
\end{aligned} \tag{4.12}
$$

where

$$
\begin{aligned}
\boldsymbol{\Theta}(i|k) = \Bigg[&\prod_{n=1}^{i-1} \boldsymbol{A_K}(n|k)\boldsymbol{E}(0|k) \quad \prod_{n=2}^{i-1} \boldsymbol{A_K}(n|k)\boldsymbol{E}(1|k) \quad \cdots \\
&\prod_{n=i-1}^{i-1} \boldsymbol{A_K}(n|k)\boldsymbol{E}(i-2|k) \quad \boldsymbol{E}(i-1|k) \Bigg]
\end{aligned} \tag{4.13}
$$

and

$$\boldsymbol{\Phi}(i|k) = \left[\Delta\boldsymbol{b}^{\mathrm{T}}(0|k) \quad \Delta\boldsymbol{b}^{\mathrm{T}}(1|k) \quad \cdots \quad \Delta\boldsymbol{b}^{\mathrm{T}}(i-1|k) \right]^{\mathrm{T}}. \tag{4.14}$$

For general compact set $\mathbb{W} = [-\Delta\boldsymbol{b}_{\max}, \Delta\boldsymbol{b}_{\max}]$ with

$$\Delta\boldsymbol{b}_{\max} = \left[\sqrt{2}\,\delta_u\,\mathrm{erf}^{-1}(p_u) \quad \sqrt{2}\,\delta_u\,\mathrm{erf}^{-1}(p_v) \quad \sqrt{2}\,\delta_u\,\mathrm{erf}^{-1}(p_r) \right]^{\mathrm{T}}$$

determined by the uncertainty distribution mean zero and standard deviation $\boldsymbol{\delta}$ and a specified probability of robustness $\boldsymbol{p} = [p_u \; p_v \; p_r]^{\mathrm{T}}$, a solution to (4.12) $\sim$ (4.14) is guaranteed to exist and each constraint is tightened for nominal states as

$$\boldsymbol{C}_j(i|k)\Delta\bar{\boldsymbol{x}}(i|k) \leqslant d_j(i|k) - d_j^0(i|k) - d_x^j(i|k). \tag{4.15}$$

For the structured norm bounded uncertainties in our case, explicit maximization [63] based on duality norm [64] is applicable to avoid solving the maximization problem (4.12) for each constraint:

$$
\begin{aligned}
d_x^j(i|k) &:= \|\boldsymbol{C}_j(i|k)\boldsymbol{\Theta}(i|k)\boldsymbol{T}(i)\|_1 = \max_{\Delta\boldsymbol{b}_t(i|k)} \\
&\left\{ \boldsymbol{C}_j(i|k)\boldsymbol{\Theta}(i|k)\boldsymbol{T}(i)\Delta\boldsymbol{b}_t(i|k) \big| \|\Delta\boldsymbol{b}_t(i|k)\|_\infty \leqslant 1 \right\},
\end{aligned} \tag{4.16}
$$

where $\Delta\boldsymbol{b}_t(i|k) = \boldsymbol{T}^{-1}(i)\Delta\boldsymbol{b}(i|k)$ and $\boldsymbol{T}(i)$ is a diagonal translation matrix

$$\boldsymbol{T}(i+1) = \begin{bmatrix} \boldsymbol{T}(i) & \boldsymbol{0} \\ \boldsymbol{0} & \mathrm{diag}(\Delta\boldsymbol{b}_{\max}) \end{bmatrix} \tag{4.17}$$

Note that the possibility of explicit maximization not only avoids solving the programming problem $(4.12) \sim (4.14)$, but also enables the formulation of an mixed-integer convex problem for our cost-effective RMPC in 4.6. Tightened constraints on nominal control inputs are handled in a similar way.

Denote the constraint sets for tightened states and control inputs as $\mathbb{C}'_x$ and $\mathbb{C}'_u$, respectively, then the following nominal optimal problem with $\mathbb{C}'_x$ and $\mathbb{C}'_u$, which has the same computational complexity with conventional MPC, is solved each sampling time:

$$(\Delta \bar{\boldsymbol{x}}^*(i|k), \Delta \bar{\boldsymbol{u}}^*(i|k)) = \underset{\Delta \bar{\boldsymbol{x}}(i|k), \Delta \bar{\boldsymbol{u}}(i|k)}{\arg \min} \tag{4.18}$$

$$\left\{ \bar{J}(\boldsymbol{x}(k), \Delta \bar{\boldsymbol{u}}(i|k)) \,|\, \Delta \bar{\boldsymbol{x}}(i|k) \in \mathbb{C}'_x, \Delta \bar{\boldsymbol{u}}(i|k) \in \mathbb{C}'_u \right\}.$$

In problem (4.10) or (4.18), system performance defined in the cost function such as path tracking, energy consumptions and arrival punctuality depends on tightened offsets which in turn depend on uncertainty distributions and the specified probability p. Large p increases safety level, i.e., the controller is robust to a larger set of uncertainties, but leads to degraded system performance and even infeasibility of online optimization or controller failures. The value of p should be set considering uncertainty properties, system capability and performance criteria systematically. An integrated design achieving this goal for problem (4.18) is proposed next.

4.6 COST-EFFECTIVE ROBUST CONTROL AGAINST STOCHASTIC MARITIME DISTURBANCES

For a standard normal distribution, the uncertainty compact set bounds increase exponentially as the probability approaching 1, and thus degrading system performance dramatically. Therefore, a practical design should increase safety levels at a cost-effective price of being robust. The idea is to integrate and make explicit use of the known probabilistic distributions relating probability and uncertainty bounds in online optimizations. However, distribution functions like the Gauss error function are usually too complex for online optimizations. A sampling based modeling approach is proposed to approximate probabilistic distributions.

We define a bound vector $\boldsymbol{Z} \in \mathbb{R}^n$ that is sampled with proper equal intervals from $[0, 3.9]^2$. Then, a corresponding probability vector $\boldsymbol{P}$ for random variable $z \sim N(0,1)$ is obtained by evaluating the Gauss error function at $\frac{\boldsymbol{Z}}{\sqrt{2}}$. Introduce an n-dimensional binary vector $\boldsymbol{a}$, then the function $z = \sqrt{2}\mathrm{erf}^{-1}(p)$ can be approximately modeled as:

$$p = \boldsymbol{a}^\mathsf{T} \boldsymbol{P}, \tag{4.19}$$

$$z = \boldsymbol{a}^\mathsf{T} \boldsymbol{Z}, \tag{4.20}$$

$$\sum \boldsymbol{a} = 1. \tag{4.21}$$

We further append a big value M and a probability 1 to $\boldsymbol{Z}$ and $\boldsymbol{P}$, respectively, to approximate the real function achieving ∞ by $p = 1, z = M, a = \begin{bmatrix} \boldsymbol{0}_{1 \times n} & 1 \end{bmatrix}^\mathsf{T}$.

---

[2]For standard normal distribution, $\mathrm{erf}(\frac{3.9}{\sqrt{2}})$=0.9999.

To this end, we have time-varying compact sets $\mathbb{W}(i|k)$ whose bounds are

$$\Delta\boldsymbol{b}_{\max}(i|k) = \boldsymbol{\delta}(i|k)\boldsymbol{a}^{\mathrm{T}}(i|k)\boldsymbol{Z}, \tag{4.22}$$

where $\boldsymbol{a}^{\mathrm{T}} = [a_u \ \ a_v \ \ a_r]^{\mathrm{T}}$ with some abuse of notation. System performance are then dependent on $\boldsymbol{a}$. Since the probability of uncertainties that the controller can be robust to is desirable to approach 1, a penalty term is added to the cost function formulating a cost-effective RMPC against stochastic uncertainties as:

$$
\begin{aligned}
(\Delta\bar{\boldsymbol{x}}^*(i|k), \Delta\bar{\boldsymbol{u}}^*(i|k), \boldsymbol{a}^*(i|k)) = &\underset{\Delta\bar{\boldsymbol{x}}(i|k),\Delta\bar{\boldsymbol{u}}(i|k),\boldsymbol{a}(i|k)}{\arg\min} \\
&\left\{ \bar{J}\left(\boldsymbol{x}(k), \Delta\bar{\boldsymbol{u}}(i|k)\right) + \boldsymbol{Q}_p \left\| 1_{3\times 1} - \boldsymbol{a}^{\mathrm{T}}(i|k)\boldsymbol{P} \right\|_1 \right. \\
&\left. |\Delta\bar{\boldsymbol{x}}(i|k) \in \mathbb{C}_x'', \Delta\bar{\boldsymbol{u}}(i|k) \in \mathbb{C}_u'', \sum \boldsymbol{a}^{\mathrm{T}}(i|k) = 1_{3\times 1} \right\},
\end{aligned} \tag{4.23}
$$

where $\sum$ performs row summations; $\boldsymbol{Q}_p$ is a weight matrix; $\mathbb{C}_x''$ and $\mathbb{C}_u''$ are the nominal state and input constraint sets using time-varying bounds (4.22).

Problem (4.23) is a mixed integer convex programming problem that can be solved efficiently. There are $3N_{\mathrm{p}}(n+1)$ binary variables involved the number of which can be reduced via several reasonable simplifications. A same probability could be applied to uncertainties in $\Delta\boldsymbol{b}_u$, $\Delta\boldsymbol{b}_v$ and $\Delta\boldsymbol{b}_r$. Furthermore, the probability could be fixed over the prediction N_{p}. Besides, a proper sampling of the probability function with a decent n could also help.

4.7 SIMULATIONS AND DISCUSSIONS

In this section, we present simulation results to demonstrate the effectiveness of the proposed robust control of USVs with stochastic uncertainties. The environmental disturbances can be roughly forecasted, but stochastic uncertainties exist in the predictions. The USV is required to track the shortest reference path between the origin and destination smoothly and arrive the destination punctually in an energy efficient and safe way despite of environmental disturbances due to wind, waves, and currents.

Controller parameters are set as follows: prediction horizon $N_{\mathrm{p}} = 6$; weight parameter $\boldsymbol{Q}_p = 10^8$; sampling time $T_s = 4$ s; stochastically uncertain predictions of environmental disturbances $\boldsymbol{b} \sim N(\bar{\boldsymbol{b}}, \boldsymbol{\Sigma})$ with mean value $\bar{\boldsymbol{b}} = \begin{bmatrix} 171500\mathrm{N} & 171500\mathrm{N} & 7203000\mathrm{Nm} \end{bmatrix}^{\mathrm{T}}$ and variances $\boldsymbol{\Sigma} = \mathrm{diag}([242538^2 \ \ 242538^2 \ \ 16977634^2])$. Physical system constraints are: $1.94\mathrm{m/s} \leqslant u \leqslant -1.11\mathrm{m/s}$ and $\boldsymbol{\tau}_{\max} = -\boldsymbol{\tau}_{\min} = \begin{bmatrix} 1372000\mathrm{N} & 1372000\mathrm{N} & 72030000\mathrm{Nm} \end{bmatrix}^{\mathrm{T}}$. Furthermore, a static obstacle is placed half way of the reference path. For simplicity, the implemented cost-effective RMPC uses a fixed probability over the prediction horizon at each sampling time.

4.7.1 Multiple simulations

For different realizations of stochastically generated uncertainties, we run 50 simulations for two controllers: a nominal MPC controller without consideration of the

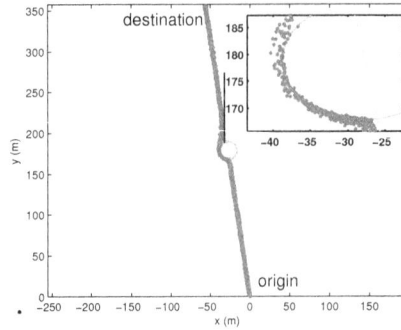

Figure 4.1: 38 trajectories of 50 simulations by nominal MPC

uncertainties in disturbance predictions and the proposed cost-effective RMPC controller.

Due to the ignorance of uncertainties in the nominal controller, 12 simulations out of 50 reports infeasibility of online optimizations. Fig. 4.1 shows the other 38 feasible trajectories when the nominal MPC controller is applied. Although the 38 simulations encounter no online infeasibility, real trajectories still fail to stay outside the obstacle area. However, the cost-effective RMPC controller reports successfully solved optimizations in all 50 simulations. Furthermore, all the trajectories successfully avoid the obstacle, as shown in Fig. 4.2. Note also that whereas in the deterministic situations, the trajectories typically are at the boundary of an obstacle, trajectories from the cost-effective RMPC are a clear distance away from the boundary of the obstacle. That is the price of robustness against the worst case of bounded uncertainties which, if stochastic, are rare to occur. It is noted that for those realizations beyond the bounds that are even rarer, the tube-based cost-effective RMPC may fail. However, any vessel with limited power cannot stand up the worst sea. The fact that infeasibility is not observed in these 50 simulations might either because those low probability realizations outside the bounds do not arise or because the designed and inherent robustness still take care of them even if they arise. When uncertainties that do cause infeasibility issues in cost-effective RMPC, rarely though, safety/recovery procedures need to be activated.

4.7.2   One simulation

The robustness of the cost-effective RMPC is due to containing the tube of all possible uncertain trajectories in the actual constraints, or equivalently, by constraining the nominal trajectory in the tightened constraints. As shown in Fig. 4.3, the nominal trajectory is a distance away from the obstacle boundaries since the obstacle avoidance constraint has been tightened in the nominal optimization problem and intuitively, forming a buffer zone outside the obstacle area. The actual trajectory which is steered to the nominal trajectory is guaranteed to stay within the tube if real uncertainties are within the bounded sets. It is the tube with the nominal trajectory as center and bounding actual uncertain trajectories that achieves both system

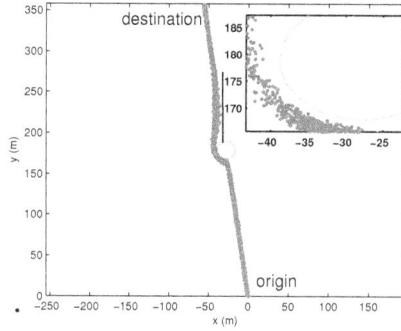

Figure 4.2: 50 trajectories of 50 simulations by cost-effective RMPC

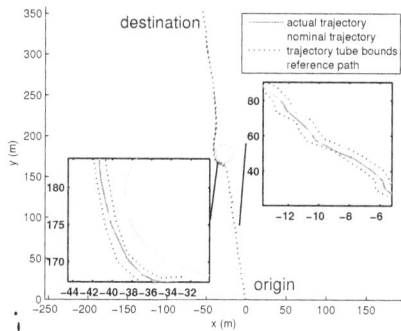

Figure 4.3: Trajectory tube of one simulation

performance (e.g., reference tracking as shown in the right magnifier in Fig. 4.3) and system safety (e.g., avoiding obstacles as shown in the left magnifier in Fig. 4.3).

The main advantages of the cost-effective RMPC over conventional tube-based RMPC are that cost-effective RMPC optimally trades-off system performance and safety level and guarantees problem solvability. This is achieved by using varying probability and thus varying uncertainty bounds as the red dots show in subplot (a) and (b) of Fig. 4.4. The deviations of the optimal probability to 1 are necessary for the feasibility of online optimizations and the deviations are as small as possible. Therefore, real uncertainties as blue squares show in subplot (b) of Fig. 4.4 are in the optimally bounded set with a high probability. Meantime, system performance in subplot (c) represented by values of the nominal cost function $\bar{J}$ is reliable. The peaks around 250 s are caused by large reference tracking errors during obstacle avoidance. However, for conventional tube-based RMPC, there is no systematic way in selecting a proper robust probability. Moreover, fixed probability causes infeasibility and thus controller failures.

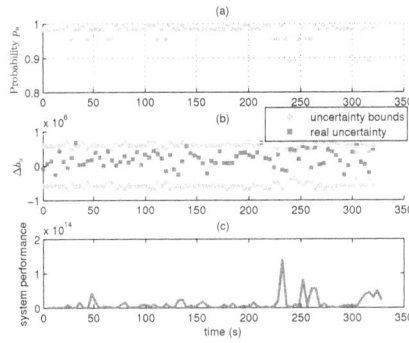

Figure 4.4: Probability, uncertainty bounds, and system performance

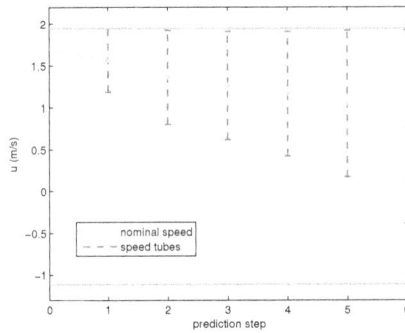

Figure 4.5: Speed tube over the prediction horizon at one sampling time

4.7.3 One sampling time over the prediction horizon

Fig. 4.5 further shows the speed tube over the prediction horizon at one sampling time. Since a fixed probability and thus fixed uncertainty bounds over the prediction at each sampling time are implemented, the effects of uncertainty propagating according to (4.7) are expected to expand but bounded over the prediction horizon as shown in Fig. 4.5. Therefore, tighter constraints are imposed to the nominal problem narrowing its feasible region, which is why the nominal speed decrease as the prediction step proceeds. However, due to the integrated design of cost-effective RMPC, the feasibility of online nominal problems are guaranteed whenever an un-tightened nominal problem is feasible.

4.8 CONCLUSIONS

In this chapter, we discuss the typical USV control problems, the mainstream control techniques, particularly the model-based control techniques. Mostly, nonlinear control approaches can provide analytical control laws with guaranteed closed-loop system properties based on USV mathematical models. Optimal control approaches can consider explicitly the system performance and constraints. However, each control method also has its potential limitations. One should choose desired control methods in specific applications.

Moreover, as an extension of the common USV control problem, an advanced cost-effective robust MPC controller for USVs is also proposed in this chapter. The proposed controller takes into account functional specifications as smooth reference tracking, arrival time awareness in an energy-efficient and safe way in spite of uncertainties due to inaccurately predicted environmental disturbances such as wind, waves, and currents. The robustness is achieved by integrating explicitly uncertainty properties, i.e., probabilistic distributions, with tube-based RMPC. This novel design yields a cost-effective solution robust to possibly infinite support stochastic uncertainties. Trade-off among safety, system performance, problem solvability, and conservativeness is achieved systematically. Future research will consider multiple USVs controlled in a distributed robust way.

Bibliography

[1] T. Fossen and T. Perez, "Kalman filtering for positioning and heading control of ships and offshore rigs," *IEEE Control Systems Magazine*, vol. 29, no. 6, pp. 32–46, 2009.

[2] D. Ma, W. Ma, S. Jin, and X. Ma, "Method for simultaneously optimizing ship route and speed with emission control areas," *Ocean Engineering*, vol. 202, p. 107170, 2020.

[3] S. Sivaraj, S. Rajendran, and L. P. Prasad, "Data driven control based on Deep Q-Network algorithm for heading control and path following of a ship in calm water and waves," *Ocean Engineering*, vol. 259, p. 111802, 2022.

[4] H. Zheng, R. R. Negenborn, and G. Lodewijks, "Predictive path following with arrival time awareness for waterborne AGVs," *Transportation Research Part C: Emerging Technologies*, vol. 70, pp. 214 – 237, 2016.

[5] H. Zheng, "Coordination of Waterborne AGVs," Ph.D. dissertation, Delft University of Technology, Delft, The Netherlands, 2016.

[6] H. Zheng, R. R. Negenborn, and G. Lodewijks, "Explicit use of probabilistic distributions in robust predictive control of waterborne AGVs — A cost-effective approach," In Proceedings of *2016 European Control Conference (ECC)*, Aalborg, Denmark, pp. 1278-1283, 2016.

[7] E. I. Sarda, I. R. Bertaska, A. Qu, and K. D. von Ellenrieder, "Development of a USV station-keeping controller," in *Proceedings of OCEANS*, Genova, Switzerland, 2015, pp. 1–10.

[8] T. I. Fossen, *Handbook of Marine Craft Hydrodynamics and Motion Control*, 2nd ed., John Wiley and Sons Ltd., West Sussex, U.K., 2021.

[9] H. Zheng, J. Wu, W. Wu, and Y. Zhang, "Robust dynamic positioning of autonomous surface vessels with tube-based model predictive control," *Ocean Engineering*, vol. 199, p. 106820, 2020.

[10] H. S. Halvorsen, H. Øveraas, O. Landstad, V. Smines, T. I. Fossen, and T. A. Johansen, "Wave motion compensation in dynamic positioning of small autonomous vessels," *Journal of Marine Science and Technology*, vol. 26, pp. 693–712, 2021.

[11] W. Yuan and X. Rui, "Deep reinforcement learning-based controller for dynamic positioning of an unmanned surface vehicle," *Computers and Electrical Engineering*, vol. 110, p. 108858, 2023.

[12] T. I. Fossen, M. Breivik, and R. Skjetne, "Line-of-sight path following of underactuated marine craft," in *Proceedings of the 6th IFAC on Manoeuvring and Control of Marine Craft*, Girona, Spain, 2003, pp. 244–249.

[13] Z. Tian, H. Zheng, and W. Xu, "Path following of autonomous surface vehicles with line-of-sight and nonlinear model predictive control," in *Proceedings of the 6th International Conference on Transportation Information and Safety*, Wuhan, China, 2021, pp. 1269–1274.

[14] Y. Zhao, X. Qi, Y. Ma, Z. Li, R. Malekian, and M. A. Sotelo, "Path following optimization for an underactuated usv using smoothly-convergent deep reinforcement learning," *IEEE Transactions on Intelligent Transportation Systems*, vol. 22, no. 10, pp. 6208–6220, 2020.

[15] D. Ma, X. Chen, W. Ma, H. Zheng, and F. Qu, "Neural network model-based reinforcement learning control for auv 3-D path following," *IEEE Transactions on Intelligent Vehicles*, vol. 9, no. 1, pp. 6208–6220, 2024.

[16] B. Piaggio, V. Garofano, S. Donnarumma, A. Alessandri, R. Negenborn, and M. Martelli, "Follow-the-leader guidance, navigation, and control of surface vessels: Design and experiments," *IEEE Journal of Oceanic Engineering*, 2023.

[17] Y. Jiang, Z. Peng, D. Wang, and C. P. Chen, "Line-of-sight target enclosing of an underactuated autonomous surface vehicle with experiment results," *IEEE Transactions on Industrial Informatics*, vol. 16, no. 2, pp. 832–841, 2019.

[18] J. Ma, J. Jin, D. Liu, L. Liu, D. Wang, and M. Chen, "An improved adaptive dynamic window algorithm for target tracking of unmanned surface vehicles," in *Proceedigns of International Conference on Signal Processing, Communications and Computing*, Xi'an, China, 2021, pp. 1–5.

[19] H. Zheng, R. R. Negenborn, and G. Lodewijks, "Trajectory tracking of autonomous vessels using model predictive control," in *Proceedings of the 19th IFAC World Congress*, Cape Town, South Africa, August 2014, pp. 8812–8818.

[20] H. Zheng, J. Li, Z. Tian, C. Liu, and W. Wu, "Hybrid physics-learning model based predictive control for trajectory tracking of unmanned surface vehicles," *IEEE Transactions on Intelligent Transportation Systems*, 2024.

[21] W. Zhou, Z. Xu, Y. Wu, J. Xiang, and Y. Li, "Energy-based trajectory tracking control of under-actuated unmanned surface vessels," *Ocean Engineering*, vol. 288, p. 116166, 2023.

[22] J. Shin, D. J. Kwak, and Y.-i. Lee, "Adaptive path-following control for an unmanned surface vessel using an identified dynamic model," *IEEE/ASME Transactions on Mechatronics*, vol. 22, no. 3, pp. 1143–1153, 2017.

[23] Z. Lu, J. Li, P. Wu, H. Zheng, and Y. Huang, "Hydrodynamic model identification of an unmanned surface vessel with experiment data," in *Proceedings of the 6th CAA International Conference on Vehicular Control and Intelligence*, Nanjing, China, 2022, pp. 1–6.

[24] H. Abrougui, S. Nejim, S. Hachicha, C. Zaoui, and H. Dallagi, "Modeling, parameter identification, guidance and control of an unmanned surface vehicle with experimental results," *Ocean Engineering*, vol. 241, p. 110038, 2021.

[25] W. B. Klinger, I. R. Bertaska, K. D. von Ellenrieder, and M. R. Dhanak, "Control of an unmanned surface vehicle with uncertain displacement and drag," *IEEE Journal of Oceanic Engineering*, vol. 42, no. 2, pp. 458–476, 2016.

[26] J. Woo, J. Park, C. Yu, and N. Kim, "Dynamic model identification of unmanned surface vehicles using deep learning network," *Applied Ocean Research*, vol. 78, pp. 123–133, 2018.

[27] R. Skjetne, "The maneuvering problem," Ph.D. dissertation, Department of Engineering Cybernetics, Norwegian University of Science and Technology, Trondheim, Norway, 2005.

[28] K. D. Do and J. Pan, "Robust path-following of underactuated ships: Theory and experiments on a model ship," *Ocean Engineering*, vol. 33, no. 10, pp. 1354–1372, 2006.

[29] Z. Peng, Y. Jiang, and J. Wang, "Event-triggered dynamic surface control of an underactuated autonomous surface vehicle for target enclosing," *IEEE Transactions on Industrial Electronics*, vol. 68, no. 4, pp. 3402–3412, 2020.

[30] W. Song, Y. Li, and S. Tong, "Fuzzy Finite-Time H_∞ Hybrid-Triggered Dynamic Positioning Control of Nonlinear Unmanned Marine Vehicles Under Cyber-Attacks," *IEEE Transactions on Intelligent Vehicles*, vol. 9, no. 1, pp. 970–980, 2024.

[31] D. Chwa, "Adaptive neural output feedback tracking control of underactuated ships against uncertainties in kinematics and system matrices," *IEEE Journal of Oceanic Engineering*, vol. 46, no. 3, pp. 720–735, 2020.

[32] L. Liu, D. Wang, Z. Peng, and Q.-L. Han, "Distributed path following of multiple under-actuated autonomous surface vehicles based on data-driven neural predictors via integral concurrent learning," *IEEE Transactions on Neural Networks and Learning Systems*, vol. 32, no. 12, pp. 5334–5344, 2021.

[33] R. Skjetne, T. I. Fossen, and P. V. Kokotovic, "Adaptive maneuvering with experiments for a model ship in a marine control laboratory," *Automatica*, vol. 41, no. 2, pp. 289–298, 2005.

[34] H. K. Khalil, *Control of Nonlinear Systems*. Prentice Hall, New York, NY, 2002.

[35] Z. Jiang, "Global tracking control of underactuated ships by Lyapunov's direct method," *Automatica*, vol. 38, no. 2, pp. 301–309, 2002.

[36] Y. Yan, X. Zhao, S. Yu, and C. Wang, "Barrier function-based adaptive neural network sliding mode control of autonomous surface vehicles," *Ocean Engineering*, vol. 238, p. 109684, 2021.

[37] A. Gonzalez-Garcia and H. Castañeda, "Guidance and control based on adaptive sliding mode strategy for a usv subject to uncertainties," *IEEE Journal of Oceanic Engineering*, vol. 46, no. 4, pp. 1144–1154, 2021.

[38] J. Rodriguez, H. Castañeda, A. Gonzalez-Garcia, and J. Gordillo, "Finite-time control for an unmanned surface vehicle based on adaptive sliding mode strategy," *Ocean Engineering*, vol. 254, p. 111255, 2022.

[39] X. Jiang, G. Xia, Z. Feng, and Z.-G. Wu, "Nonfragile formation seeking of unmanned surface vehicles: A sliding mode control approach," *IEEE Transactions on Network Science and Engineering*, vol. 9, no. 2, pp. 431–444, 2021.

[40] W. Zhou, Y. Wang, C. K. Ahn, J. Cheng, and C. Chen, "Adaptive fuzzy backstepping-based formation control of unmanned surface vehicles with unknown model nonlinearity and actuator saturation," *IEEE Transactions on Vehicular Technology*, vol. 69, no. 12, pp. 14 749–14 764, 2020.

[41] S. J. Ohrem, H. B. Amundsen, W. Caharija, and C. Holden, "Robust adaptive backstepping dp control of rovs," *Control Engineering Practice*, vol. 127, p. 105282, 2022.

[42] W. Naeem, T. Xu, R. Sutton, and A. Tiano, "The design of a navigation, guidance, and control system for an unmanned surface vehicle for environmental monitoring," *Proceedings of the Institution of Mechanical Engineers, Part M: Journal of Engineering for the Maritime Environment*, vol. 222, no. 2, pp. 67–79, 2008.

[43] H. P. Geering, *Optimal Control with Engineering Applications*. Berlin, German: Springer, 2007, vol. 113.

[44] H. Wei and Y. Shi, "MPC-based motion planning and control enables smarter and safer autonomous marine vehicles: Perspectives and a tutorial survey," *IEEE/CAA Journal of Automatica Sinica*, vol. 10, no. 1, pp. 8–24, 2023.

[45] X. Wang, Z. Zou, T. Li, and W. Luo, "Path following control of underactuated ships based on nonswitch analytic model predictive control," *Journal of Control Theory and Applications*, vol. 8, no. 4, pp. 429–434, 2010.

[46] N. Wang, Y. Gao, Y. Liu, and K. Li, "Self-learning-based optimal tracking control of an unmanned surface vehicle with pose and velocity constraints," *International Journal of Robust and Nonlinear Control*, vol. 32, no. 5, pp. 2950–2968, 2022.

[47] A. Gonzalez-Garcia, I. Collado-Gonzalez, R. Cuan-Urquizo, C. Sotelo, D. Sotelo, and H. Castañeda, "Path-following and lidar-based obstacle avoidance via nmpc for an autonomous surface vehicle," *Ocean Engineering*, vol. 266, p. 112900, 2022.

[48] A. Tsolakis, R. R. Negenborn, V. Reppa, and L. Ferranti, "Model Predictive Trajectory Optimization and Control for Autonomous Surface Vessels Considering Traffic Rules," *IEEE Transactions on Intelligent Transportation Systems*, 2024.

[49] Z. Peng, B. Zhang, O. Sun, D. Wang, M. Han, L. Liu, and H. Wang, "Finite-set model predictive speed and heading control of autonomous surface vehicles with unmeasured states," in *Proceedings of Tenth International Conference on Intelligent Control and Information Processing*, Marrakesh, Morocco, 2019, pp. 158–163.

[50] D. Q. Mayne, "Model predictive control: Recent developments and future promise," *Automatica*, vol. 50, no. 12, pp. 2967–2986, 2014.

[51] S. V. Raković, "Set theoretic methods in model predictive control," in *Nonlinear Model Predictive Control*, D. M. Raimondo, F. Allgöwer, Eds., pp. 41–54, Springer, 2009.

[52] P. O. M. Scokaert and D. Q. Mayne, "Min-max feedback model predictive control for constrained linear systems," *IEEE Transactions on Automatic Control*, vol. 43, no. 8, pp. 1136–1142, 1998.

[53] J. Löfberg, "Approximations of closed-loop minimax MPC," in *Proceedings of 42nd IEEE Conference on Decision and Control*, vol. 2, Maui, USA, 2003, pp. 1438–1442.

[54] P. J. Goulart, E. C. Kerrigan, and J. M. Maciejowski, "Optimization over state feedback policies for robust control with constraints," *Automatica*, vol. 42, no. 4, pp. 523–533, 2006.

[55] S. V. Raković, B. Kouvaritakis, M. Cannon, C. Panos, and R. Findeisen, "Parameterized tube model predictive control," *IEEE Transactions on Automatic Control*, vol. 57, no. 11, pp. 2746–2761, 2012.

[56] D. Q. Mayne, M. M. Seron, and S. V. Raković, "Robust model predictive control of constrained linear systems with bounded disturbances," *Automatica*, vol. 41, no. 2, pp. 219–224, 2005.

[57] J. A. Primbs and C. H. Sung, "Stochastic receding horizon control of constrained linear systems with state and control multiplicative noise," *IEEE Transactions on Automatic Control*, vol. 54, no. 2, pp. 221–230, 2009.

[58] B. Kouvaritakis, M. Cannon, S. V. Raković, and Q. Cheng, "Explicit use of probabilistic distributions in linear predictive control," *Automatica*, vol. 46, no. 10, pp. 1719–1724, 2010.

[59] K. Margellos, P. Goulart, and J. Lygeros, "On the road between robust optimization and the scenario approach for chance constrained optimization problems," *IEEE Transactions on Automatic Control*, vol. 59, no. 8, pp. 2258–2263, 2014.

[60] P. Hokayem, D. Chatterjee, and J. Lygeros, "On stochastic receding horizon control with bounded control inputs," in *Proceedings of the 48th IEEE Conference on Decision and Control, held jointly with the 28th Chinese Control Conference*, Shanghai, China, 2009, pp. 6359–6364.

[61] L. C. Andrews, *Special Functions for Engineers and Applied Mathematicians*, Macmillan, London, UK, 1985.

[62] S. Boyd, "Linear quadratic regulator: Discrete-time finite horizon," in *Lecture Notes: Introduction to Linear Dynamical Systems*, Stanford, USA, 2009.

[63] J. Löfberg, "Automatic robust convex programming," *Optimization Methods and Software*, vol. 27, no. 1, pp. 115–129, 2012.

[64] S. Boyd and L. Vandenberghe, *Convex Optimization*, Cambridge University Press, Cambridge, UK, 2004.

CHAPTER 5

Learning-Based Control of USVs

5.1 INTRODUCTION

MODEL based control, e.g., MPC in the previous chapter, is widely used in USV control because it can explicitly deal with nonlinear hydrodynamics and constraints conveniently. However, the practical applications of model-based methods to USVs are still rarely seen.

On the one hand, the performance of model-based control method heavily relies on the accuracy of the system model. Disregarding modeling errors may result in poor control performance, violations of constraints, or even actuator damages. On the other hand, there is usually a trade-off between the model accuracy and computational burden or controller design complexity. Moreover, due to the universal marine environmental disturbances and the uncertain USV hydrodynamics, it is in general difficult to obtain an accurate USV model. Even if highly accurate USV hydrodynamic model is readily available, the heavy computational burden of model-based computations would not meet the requirements for real-time applications.

5.2 MODEL-FREE USV CONTROL METHODS

Mostly, the mathematical USV hydrodynamics model in the complex uncertain maritime environment is difficult to obtain. If there is no USV mathematical model readily available, the controller design necessarily relies on trials and errors (e.g., proportional-integral-derivative (PID) control), human experiences (e.g., fuzzy logic control), or data-driven approaches.

The first recognized and most widely implemented controller until now is still PID [1] which was first proposed for the ship steering control. PID has the advantages of being simple to implement and low cost [2]. However, issues such as nonlinearity handling, parameter tuning, overshoots, constraints, and performance guarantee are recognized in PID design. In general, there are only limited guidelines on how to tune the PID parameters. The well-tuned parameters in one scenario might not perform well in other scenarios due to the inherent nonlinear and time-varying USV hydrodynamics. Therefore, the PID controllers are also frequently combined with other techniques to be adaptive, e.g., the fuzzy-PID [3] and the gain scheduling PID optimized by the genetic algorithm and fuzzy logic [4]. Nevertheless, it has been shown in [5] that the modeled based LQG controller still has superior performance than the model free adaptive PID. The fuzzy logic control [6] is also a cost-effective design for USVs. However, the fuzzy parameter settings heavily rely on human experiences and the system performance cannot be guaranteed.

If sufficient system data is available, the control can also be achieved with data-driven approaches by learning the system patterns from the data [7]. More than three decades ago, a single neural network already showed comparable performance to a set of optimal guidance control laws computed for different forward speeds [8] and robustness in various scenarios [9]. Until now, various learning-based control methodologies have been developed for USVs which can be divided mainly into two categories. The first category is the machine learning which includes the Gaussian Process (GP) [10] and Bayesian learning [11]. The second category is deep learning [12] which includes reinforcement learning (RL) [13][14] and Deep Neural Network (DNN) [15].

Three GP models are utilized in [16] to deal with nonlinear systems with additive state-dependent uncertainties. Sparse GPs are further designed to improve the computational efficiency. It is known that as the data accumulates, the prediction speed of the GP decreases significantly. In addition, GP requires the assumption of known Gaussion noises. Therefore, deep learning-based control receives widely attention due to its excellent learning capacity for complex USV control problems with massive amounts of data. To deal with the time-varying USV dynamics and environmental disturbances, an online DNN and an extended state observer are constructed to learn the USV dynamics and estimated learning errors respectively in [17]. The kinematic and kinetic uncertainties are handled by an adaptive neural uncertainty observer in [18].

Due to the impressive empirical performance, RL-based control has also received extensive attention. RL has been used in [19] and [20] to achieve human championship-level performance in drone racing and fast motion of legged robots, respectively. For USVs, [21] [22] realize accurate and efficient dynamic positioning, path following, and formation control through RL, respectively. Compared with the traditional control methods such as PID, it is demonstrated in [24] with sea trials that the RL-based control has faster and more accurate control performance for the USV path tracking task. The RL-based control combined with a neural network model is applied to autonomous underwater vehicles in [13], and satisfactory 3-D path following results are achieved. However, RL requires huge number of training data before being applied to the system, which brings trouble in the practical applications. Mostly, current learning-based USV control works have limited to simulations. Further practical applications especially in safety critical scenarios still need to address issues such as data availability, explain-ability and provable performance guarantees.

5.3 HYBRID USV CONTROL METHODS

Mathematical models embed the prior knowledge of the system, e.g., the hydrodynamic relationships between the fluid and the rigid USV hull. Learning approaches possess the flexibility especially in handling system uncertainties due to, e.g., wind, waves, and currents. Therefore, it is natural to combine a system nominal models with the learning powered ideas.

Generally, there are two ways in incorporating the learning system in the conventional control. One way is to build a hybrid nominal and learning system model, and then the aforementioned model-based control approaches can be applied. Since the learning models can approximate nonlinear systems with arbitrary accuracy, the model-based control performance with hybrid models could be improved. The combination of nominal first-principles models and learning models has seen applications in robotic arms [25], motor racing [26], unmanned aerial vehicles [27, 28] and also USVs [29, 30]. The GP, neural ordinary differential equations, neural network and DNN learning models have been used in [25], [27], [29] and [30], respectively, to improve the accuracy of the nominal system models. In most cases, simulation results have been presented to demonstrate the effectiveness of the hybrid modeling and control frameworks. Notably, in extreme situation such as the racing car [25], experiment results are available to show that the racing car's control performance can be improved by approximately 10%. However, due to the black-box feature of the learning models, the feedback control law based on the hybrid model usually require heavy tuning of the controller parameters to achieve satisfactory results.

The other hybrid way is to combine the conventional control action and the learning control action directly [31]. In such cases, the overall system controller is consisted of additional two parts: a baseline controller and a learning controller, e.g., the backstepping control law with RL control in [32]. On the one hand, due to the learning controller, the overall controller can usually handle system uncertainties better [32]. On the other hand, due to the conventional controller, provable closed-loop properties can usually be guaranteed under specific conditions [31]. However, the training data quality is crucial in the success of such end-to-end learning control techniques. Besides the learning-based control, hybrid fuzzy and backstepping controller are also applied to USV formation problems considering actuator saturation and unknown nonlinear items [33]. The data-driven idea is combined with the conventional PID and backstepping controllers for USVs [34] to handle the uncertainties and unknown parametric dynamics. The overall USV distributed control problem is divided hierarchically into kinematic consensus control and the kinetic integral concurrent learning control [35]. For complicated USV control problems, combinations of multiple controllers show superiority in the control performance.

Combining the model-based control methods in Chapter 4, a categorization of the main USV control methods are summarized in Table 5.1.

Considering the potential advantages of the hybrid physics and learning control methods, we next propose a new neural-sailing concept with a hybrid Physics-learning Model-based Predictive Control (PL-MPC) approach to deal with the modeling error issue in USV motion control problems. Compared with the conventional MPC,

TABLE 5.1 An overview of the main USV control methods.

	Control techniques	References	Main features
Model based control methods	Back-stepping	[33][36][37][38][39][40]	Recursive design procedure, closed-loop properties
	Sliding mode control	[29][41][43][44][45]	Chattering phenomenon, closed-loop properties
	Optimal control	[46][47][48]	Optimal performance indices, model dependence
	MPC	[49][50][51][52][53][55][56][57][58][59]	Handing constraints, quasi-optimality, online optimization
Model free control methods	PID	[1][60]	Simplicity, tedious parameter tuning in varying scenarios
	Fuzzy control	[3][4][6][61]	Human experience, no rigorous guarantee
	RL control	[13][17][21][22][23][24][62]	Impressive empirical performance, data dependence, and tedious parameter tuning
	Neural control	[7][8][9][63]	Black-box approach, non-explainable
Hybrid control methods	Hybrid modeling	[17][29][30][64][65]	Physics informed nominal modeling plus learning-based modeling
	Hybrid control	[31][32][33][34][35][42][66]	Adaptiveness to uncertainties or disturbances

PL-MPC is robust in that it handles the modeling errors caused by either the USV uncertain hydrodynamics or the environmental disturbances. The method continuously learns the modeling errors using the Deep Neural Network (DNN) due to its capability of approximating arbitrary nonlinear functions [67]. Then, the physics model and the learning model are combined to form a hybrid prediction model in MPC. Finally, with the quasi-infinite nonlinear MPC scheme, we investigate and ensure the closed-loop stability of the proposed PL-MPC method. In this way, the advantages of the flexibility of deep learning and the closed-loop properties of MPC are integrated so that greater control performance is achieved than by MPC individually. In addition, to achieve the trade-off between the modeling accuracy and computational efficiency, we further propose a successive linearization method for PL-MPC for the first time.

5.4 HYBRID PHYSICS AND LEARNING BASED ON USV MODELING

The body-fixed reference frame $x_b - y_b - o_b$ is fixed to the inertial earth-fixed frame $X - Y - O$ with the center of gravity of the USV as the center, as shown in Fig. 5.1. In the earth-fixed frame, X and Y represent the true north and true east, respectively, and ψ represents the heading angle. Generally, a USV experiences motions in six degrees of freedom (DOF). For tracking problems of surface vehicles, models with three DOFs in the horizontal plane considering surge, sway and yaw are sufficient to capture the main system characteristics [68] based on the assumption that the roll and pitch motions are small.

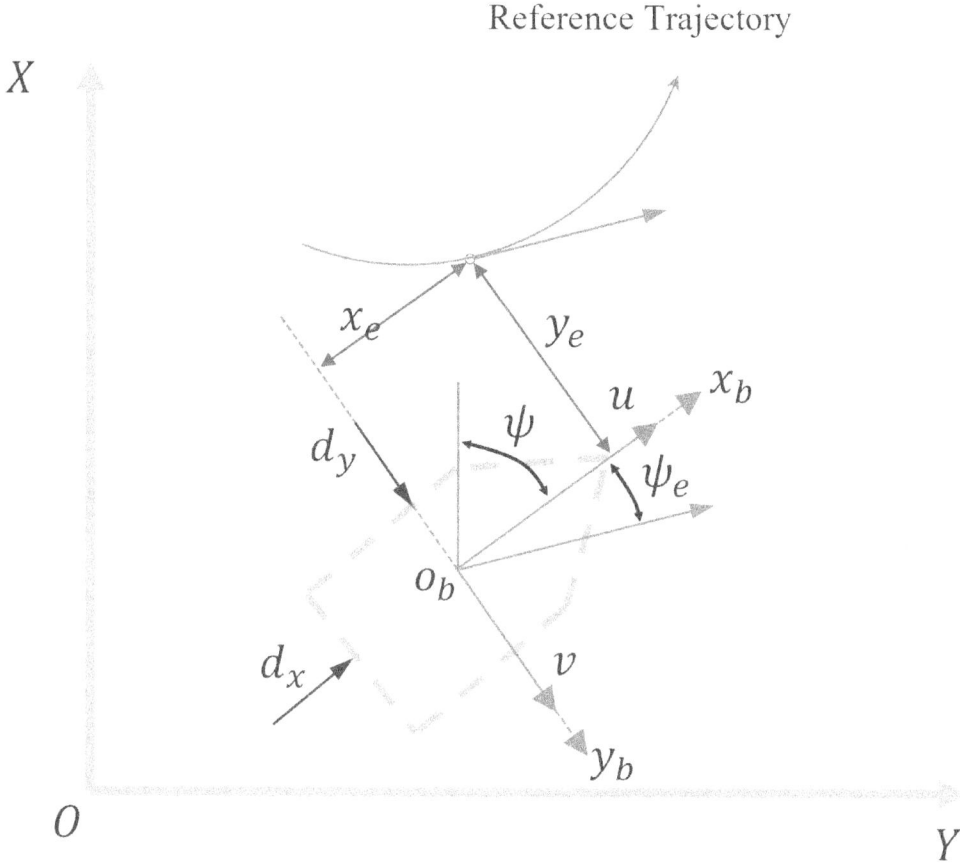

Figure 5.1: Horizontal dynamics of the USV and the trajectory tracking problem.

The velocity vector of the body-fixed reference frame of the USV is defined as $V = [u, v, r]^T$, representing the surge, sway and yaw velocities of the USV, respectively. The position and yaw angle of the USV in the inertial coordinate system are defined as $\eta = [x, y, \psi]^T$. In Fig. 5.1, x_e, y_e, and ψ_e are the error position variables in the body-fixed reference frame as to be defined in later; b_x and b_y are the unknown environmental disturbance forces.

Note that the dependence of these states on the time has been omitted whenever confusion is not caused. Thus, the three degree of freedom kinematics model of the USV is obtained as:

$$\dot{\eta} = R(\psi)V, \tag{5.1}$$

where $R(\psi)$ is the rotation matrix defined as:

$$R(\psi) = \begin{bmatrix} \cos(\psi) & -\sin(\psi) & 0 \\ \sin(\psi) & \cos(\psi) & 0 \\ 0 & 0 & 1 \end{bmatrix}.$$

The dynamics of an USV can be modeled following [68] as:

$$\begin{aligned} \dot{u} &= a_{11}vr + a_{12}r^2 + a_{13}u + a_{14}\tau_u + d_1, \\ \dot{v} &= a_{21}uv + a_{22}ur + a_{23}v + a_{24}r + a_{25}\tau_v + a_{26}\tau_r + d_2, \\ \dot{r} &= a_{31}uv + a_{32}ur + a_{33}v + a_{34}r + a_{35}\tau_v + a_{36}\tau_r + d_3, \end{aligned} \tag{5.2}$$

where $\tau = [\tau_u, \tau_v, \tau_r]^T$ is the control force and moment of the USV. The involved parameters $a_{..}$ and $d_.$ are detailed in Appendix A.2. The real dynamics model (5.2), is divided into two parts: the first part is the easily pre-obtained physical model, and the other part is the uncertainty model $d = [d_1, d_2, d_3]^T$ that is difficult to model. The uncertainty model lumps the effects due to the hydrodynamic disturbances and environmental disturbances (b_x, b_y). Therefore, DNN can handle uncertain disturbances by learning from the disturbances datasets D.

For notation simplicity, generalize the kinematics and dynamics (5.1)-(5.2) of USVs as:

$$\dot{x} = f(x, \tau), \tag{5.3}$$

where $f : \Re^6 \times \Re^3 \times \Re^3 \to \Re^6$ is a nonlinear smooth function with system states $x = [\eta^T, V^T]^T$.

Due to USV uncertain hydrodynamics, the effects of environmental disturbances and the limitations of the system identification techniques, the USV true system dynamics $f(\cdot)$, in general, cannot be accurately known. Therefore, we separate $f(\cdot)$ into two parts, i.e., the physics model that is obtained through parameter identification, and the learning model that is obtained through DNN. The two parts of models are introduced in more detail next.

5.4.1 Physics model with parameter identification

One of the main challenges for the practical applications of MPC to USVs is that the system model is difficult to obtain. The USV system involves a large number of hydrodynamic parameters and complicated couplings among these parameters, as can be seen in (5.1)-(5.2). Standard parameter identification techniques, e.g., the least squares [69], can only obtain a partial physics model $f_p(\cdot)$ from the motion data. Particularly, denote the identified physical dynamics as:

$$\begin{aligned} \dot{u} &= \hat{a}_{11}vr + \hat{a}_{12}r^2 + \hat{a}_{13}u + \hat{a}_{14}\tau_u, \\ \dot{v} &= \hat{a}_{21}uv + \hat{a}_{22}ur + \hat{a}_{23}v + \hat{a}_{24}r + \hat{a}_{25}\tau_v + \hat{a}_{26}\tau_r, \\ \dot{r} &= \hat{a}_{31}uv + \hat{a}_{32}ur + \hat{a}_{33}v + \hat{a}_{34}r + \hat{a}_{35}\tau_v + \hat{a}_{36}\tau_r, \end{aligned} \tag{5.4}$$

where we define

$$\hat{\boldsymbol{\theta}} = \begin{bmatrix} \hat{\boldsymbol{\theta}}_1 & \mathbf{0}_{1\times6} & \mathbf{0}_{1\times6} \\ \mathbf{0}_{1\times4} & \hat{\boldsymbol{\theta}}_2 & \mathbf{0}_{1\times6} \\ \mathbf{0}_{1\times4} & \mathbf{0}_{1\times6} & \hat{\boldsymbol{\theta}}_3 \end{bmatrix}$$

and $\hat{\boldsymbol{\theta}}_1 = [\hat{a}_{11}, \hat{a}_{12}, ..., \hat{a}_{14}]$, $\hat{\boldsymbol{\theta}}_2 = [\hat{a}_{21}, \hat{a}_{22}, ..., \hat{a}_{26}]$, $\hat{\boldsymbol{\theta}}_3 = [\hat{a}_{31}, \hat{a}_{32}, ..., \hat{a}_{36}]$, as the parameters that need to be identified. The USV motion data $\boldsymbol{\phi} = [\boldsymbol{\phi}_1, \boldsymbol{\phi}_2, \boldsymbol{\phi}_3]^T$ that is used for the parameter identification is $\boldsymbol{\phi}_1 = [vr, r^2, u, \tau_u]$, $\boldsymbol{\phi}_2 = [uv, ur, v, r, \tau_v, \tau_r]$, $\boldsymbol{\phi}_3 = [uv, ur, v, r, \tau_v, \tau_r]$.

The observations in the identification process is $\boldsymbol{Y} = [\dot{u}, \dot{v}, \dot{r}]^T$. Then, the error function for the parameter identification is defined as

$$e = \frac{1}{2} \left(\boldsymbol{Y} - \hat{\boldsymbol{\theta}}\boldsymbol{\phi} \right)^2,$$

which is minimized by taking the derivative with respect to $\hat{\boldsymbol{\theta}}$ and setting the derivative as zero:

$$\frac{\partial e}{\partial \hat{\boldsymbol{\theta}}} = \boldsymbol{Y}\boldsymbol{\phi}^T - \hat{\boldsymbol{\theta}}^T \boldsymbol{\phi}\boldsymbol{\phi}^T = 0. \tag{5.5}$$

Then, the physics model parameters $\hat{\boldsymbol{\theta}}$ can be identified. Analytically, we get identified physics model parameters $\hat{\boldsymbol{\theta}} = (\boldsymbol{\phi}\boldsymbol{\phi}^T)^{-1}\boldsymbol{\phi}\boldsymbol{Y}^T$.

5.4.2 Learning model with DNN

Mostly, the identified model cannot capture the real system dynamics. It is especially the case for USVs. The unmodeled dynamics $\boldsymbol{d} = [d_1, d_2, d_3]^T$ and the parameter identification errors between $f_p(\cdot)$ and $f(\cdot)$ are represented as $f_l(\cdot)$. To leverage the capability of approximating arbitrary nonlinear function of neural networks, we propose using DNN to obtain $f_l(\cdot)$. Note that although convolutional neural networks (CNNs) also have outstanding ability in terms of nonlinear function approximation, the convolutional and pooling layers of CNNs introduce significant challenges to the Taylor expansion which is important in the latter controller design. DNN shows more convenience on this aspect.

The proposed DNN structure is a feed-forward, artificial neural network with three hidden layers between its input layer and output layer, as shown in Fig. 5.2. For each hidden layer i, a nonlinear neuron *tanh* activation function is used to map all the lower layer states $\boldsymbol{l}_{i-1}$ to the upper layer states $\boldsymbol{l}_i$ as

$$\begin{aligned} \boldsymbol{l}_i &= \tanh\left(\boldsymbol{l}_{i-1}\right)\mathbf{W}_i^T + b_i, \quad \text{for } i = 2, .., m \\ \boldsymbol{l}_i &= [l_i^1, ..., l_i^j], \quad \text{for } j = 1, .., n \end{aligned} \tag{5.6}$$

where m is the number of the hidden layers; l_i^j is the output of the neuron j in the layer i; n is the number of neurons in each hidden layer; b_i represents the bias of layer i; $\mathbf{W}_i$ is the connection weights between layer i and layer $i-1$. Specially, $\boldsymbol{l}_0$ represents the input layer of the network. The input to the neural network is $[\boldsymbol{V}, \boldsymbol{\tau}]$. The outputs of the neural network are the modeling errors $\boldsymbol{d}$. As for the activation function, *tanh* is a differentiable function and its derivative is bounded. In addition, the output range of *tanh* is between (-1, 1) which is consistent with the distribution of standardized training data. Therefore, *tanh* is set as the activation function.

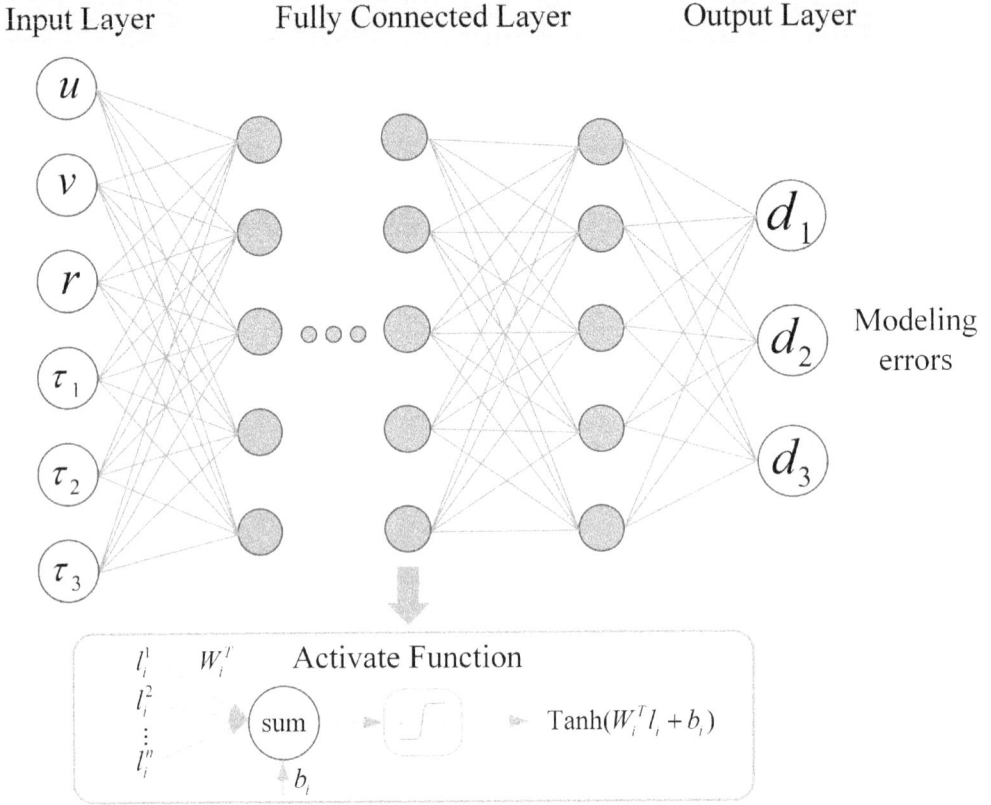

Figure 5.2: The DNN structure to learn the USV modeling errors.

DNN training data: To obtain the data to train the DNN, we conduct multiple simulations using the complete USV model (5.1)-(5.2), and using the identified physics model (5.1) and (5.4). Specifically for the USV trajectory tracking problem, in each trial (i.e., one complete trajectory tracking simulation) j, denote the obtained data set as $\boldsymbol{D}^j = \{\boldsymbol{a}^j(k)\}$ with k being the discrete time steps and $k = 1, 2, \cdots$. The data obtained per time step is $\boldsymbol{a}^j(k) = ([\boldsymbol{V}^j(k), \boldsymbol{\tau}^j(k)], \boldsymbol{d}^j(k))$. Moreover, to improve the DNN's learning performance, the training data is dealt with the standardization process. Specifically, at time k of the j-th trial, the standardized training data $\boldsymbol{a}_y^j(k)$ is calculated as: $\boldsymbol{a}_y^j(k) = (\boldsymbol{a}^j(k) - \bar{\boldsymbol{a}})/\boldsymbol{\sigma}_a$, where $\bar{\boldsymbol{a}}$ is the mean of the training data, and $\boldsymbol{\sigma}_a$ is the standard deviation of training data. With the training of the DNN, we get the learning model $f_l(\boldsymbol{V}, \boldsymbol{\tau})$ that captures the modeling error dynamics.

To this end, we are able to represent the USV dynamics with an easily identified physics model $f_p(\boldsymbol{x}, \boldsymbol{\tau})$ and the error learning model $f_l(\boldsymbol{V}, \boldsymbol{\tau})$, i.e.,

$$f(\boldsymbol{x}, \boldsymbol{\tau}) = f_p(\boldsymbol{x}, \boldsymbol{\tau}) + f_l(\boldsymbol{V}, \boldsymbol{\tau}), \tag{5.7}$$

where $f_p(\boldsymbol{x}, \boldsymbol{\tau})$ and $f_l(\boldsymbol{V}, \boldsymbol{\tau})$ instead of the hardly available $f(\boldsymbol{x}, \boldsymbol{\tau})$ will be used as the prediction models in the following MPC design.

5.5 HYBRID PHYSICS-LEARNING MODEL-BASED PREDICTIVE CONTROL – PL-MPC

In this section, with the identified model in Section 5.4.1 and the learned model in Section 5.4.2, we propose the hybrid physics-learning model-based predictive control, i.e., PL-MPC, for the USV trajectory tracking problem. Moreover, the closed-loop asymptotic stability of PL-MPC is also analyzed. To achieve a trade-off between the prediction accuracy and computational efficiency, we apply the successive linearization technique as also introduced approach in the previous chapter for the proposed PL-MPC.

5.5.1 Error dynamics with the hybrid model

As shown in Fig. 5.1, the trajectory tracking task requires the USV to track to the reference trajectory which is a function of time. Since the reference positions are time-varying, we first convert the trajectory tracking control problem into a stabilization problem by introducing the error dynamics model. In the new error reference framework, the PL-MPC controller is designed and the closed-loop stability is analyzed.

Denote the desired system states as $\boldsymbol{x}_d = [x_d, y_d, \psi_d, u_d, v_d, r_d]^T$ and the error states as $\boldsymbol{x}_e = [x_e, y_e, \psi_e, u_e, v_e, r_e]^T$, then the error states can be derived in the USV parallel reference frame [70] as:

$$
\begin{aligned}
x_e &= \cos\left(\psi\right)\left(x_d - x\right) + \sin\left(\psi\right)\left(y_d - y\right), \\
y_e &= -\sin\left(\psi\right)\left(x_d - x\right) + \cos\left(\psi\right)\left(y_d - y\right), \\
\psi_e &= \psi_d - \psi, \\
u_e &= u - \left(u_d \cos\left(\psi_e\right) - v_d \sin\left(\psi_e\right)\right), \\
v_e &= v - \left(u_d \sin\left(\psi_e\right) + v_d \cos\left(\psi_e\right)\right), \\
r_e &= r - r_d.
\end{aligned}
\tag{5.8}
$$

By differentiating both sides of the equations (5.8) with respect to time, we have

$$
\begin{bmatrix}
\dot{x}_e \\
\dot{y}_e \\
\dot{\psi}_e \\
\dot{u}_e \\
\dot{v}_e \\
\dot{r}_e
\end{bmatrix}
=
\begin{bmatrix}
y_e r_d - u_e + y_e r_e \\
-x_e r_d - v_e - x_e r_e \\
-r_e \\
h_e\left(\boldsymbol{x}, \boldsymbol{x}_d, \boldsymbol{\tau}\right)
\end{bmatrix},
\tag{5.9}
$$

where

$$
h_e\left(\boldsymbol{x}, \boldsymbol{x}_d, \boldsymbol{\tau}\right) = f_p\left(\boldsymbol{x}, \boldsymbol{\tau}\right) + f_l\left(\boldsymbol{V}, \boldsymbol{\tau}\right) + f_d\left(\boldsymbol{x}_e, \boldsymbol{x}_d\right),
\tag{5.10}
$$

with $f_d = \begin{bmatrix} -u_d \sin\left(\psi_e\right) r_e - v_d \cos\left(\psi_e\right) r_e \\ u_d \cos\left(\psi_e\right) r_e - v_d \sin\left(\psi_e\right) r_e \\ 0 \end{bmatrix}$. The functions f_p and f_l are defined

as in Section 5.4.1 and Section 5.4.2, respectively. Therefore, we have obtained the hybrid physics-learning tracking error model (5.9). This hybrid error model will be used as the prediction model in the next to design an MPC controller for the USV trajectory tracking.

5.5.2 PL-MPC for USV trajectory tracking

MPC computes a sequence of control inputs that optimize the system behavior over a receding horizon based on the prediction model and the current state [49]. The first input of the sequence is applied to the system. The entire process is repeated at the next MPC step with the new system state.

With the hybrid tracking error model (5.9), the USV trajectory tracking problem is converted to a stabilization problem. The control objective is to steer the error states $\boldsymbol{x}_e$ to the origin, which means the USV trajectory approaches the desired trajectory. Setting the prediction horizon as N_p, then the PL-MPC problem, is formulated as follows:

$$\min_{\boldsymbol{\tau}} J\left(\boldsymbol{x}_e, \Delta\boldsymbol{\tau}\right) = \sum_{i=0}^{N_p-1} \ell\left(\boldsymbol{x}_e(i \mid k), \Delta\boldsymbol{\tau}(i \mid k)\right) + F\left(\boldsymbol{x}_e(N_p \mid k)\right)$$

$$\begin{aligned}
\text{subject to} \quad & (5.9), \\
& \boldsymbol{x}_e(i+1 \mid k) \in \Omega_{x_e}, \\
& \boldsymbol{x}_e(N_p \mid k) \in \Omega, \\
& \boldsymbol{\tau}(i \mid k) \in \Omega_{\tau},
\end{aligned} \tag{5.11}$$

for $i = 0, 1, \ldots, N_p - 1$ and $(i|k)$ represents the i-th prediction step at time step k. In (5.11),

$$\ell\left(\boldsymbol{x}_e(i \mid k), \; \boldsymbol{\tau}(i \mid k)\right) = \boldsymbol{x}_e^T(i \mid k)Q_f\boldsymbol{x}_e(i \mid k) + \Delta\boldsymbol{\tau}^T(i \mid k)\mathcal{R}\Delta\boldsymbol{\tau}(i \mid k) \tag{5.12}$$

is the stage cost, and $F\left(\boldsymbol{x}_e(N_p \mid k)\right) = \boldsymbol{x}_e^T(N_p \mid k)\mathcal{Q}_f\boldsymbol{x}_e(N_p \mid k)$ is the terminal cost. $Q_f \in \Re^{6\times6}$ and $\mathcal{R} \in \Re^{3\times3}$ denote positive-definite weighting matrices, and $\mathcal{Q}_f \in \Re^{6\times6}$ is the terminal weighting matrix. Ω_x denotes the error state constraint set, and Ω_{τ} denotes the control constraint set. Ω is a designed terminal region encompassing the origin, in which the predicted terminal state $\boldsymbol{x}_e(N_p \mid k)$ will end up to guarantee the closed-loop stability.

5.5.3 Closed-loop asymptotic stability of PL-MPC

From a theoretical point of view, an infinite prediction horizon $N_p = \infty$ ensures the closed-loop stability of PL-MPC. For linear MPC with finite prediction horizon, terminal cost and constraints are usually adopted to ensure the stability of the closed-loop system. In [71], the authors propose an method deriving the terminal cost and constraints for nonlinear systems. A conservative elliptical terminal region is designed to ensure the stability, in which the feedback controller of the linearized system can substitute the nonlinear one. If the linearization error is small, we can assume that the linearized controller can approximate the nonlinear one in the whole experiment region. Thus we can extend the terminal region with the help of the maximum invariant set. In this way, large prediction horizons are not necessary to avoid heavy computational burden. For the proposed PL-MPC, we follow the procedures in [52] to calculate the terminal constraints and cost as:

Step 1: Linearizing the USV error dynamics (5.9) at the origin, then we have:

$$\dot{\boldsymbol{x}}_e = \mathbf{A}\boldsymbol{x}_e + \mathbf{B}\boldsymbol{\tau} \tag{5.13}$$

where $\mathbf{A} = (\partial f_e/\partial \boldsymbol{x}_e)|_{\boldsymbol{x}_e=0}$ and $\mathbf{B} = (\partial f_e/\partial \boldsymbol{\tau})|_{\boldsymbol{\tau}=0}$. If (5.13) is stabilizable, we can derive the state feedback control law $\boldsymbol{\tau} = K\boldsymbol{x}_e$.

Step 2: Chose a $\kappa \in [0, \infty)$ satisfying $\kappa < -\lambda_{\max}(\mathbf{A}_K)$ where $\mathbf{A}_K = \mathbf{A} + \mathbf{B}K$. Solve the Lyapunov equation $(\mathbf{A}_K + \kappa I)^\top \mathcal{Q}_f + \mathcal{Q}_f(\mathbf{A}_K + \kappa I) = -Q_f^\star$ to get a positive-definite and symmetric terminal weighting matrix $\mathcal{Q}_f$.

Step 3: Calculate the maximum invariant set Ω of the linearized system (5.13) as in [52].

According to [71], the closed-loop asymptotic stability of a nonlinear system with nonlinear MPC is ensured when the following four assumptions are satisfied :

A1: $\Omega \subset \Omega_x$, and Ω is closed, $0 \in \Omega$.

A2: $K\boldsymbol{x}_e \in \Omega_\tau, \forall \boldsymbol{x} \in \Omega$.

A3: Ω is positively invariant under K.

A4: $F(\mathbf{A}\boldsymbol{x}_e + \mathbf{B}\boldsymbol{\tau}) - F(\boldsymbol{x}_e) + \ell(\boldsymbol{x}_e, \boldsymbol{\tau}) \leq 0, \forall \boldsymbol{x} \in \Omega$.

For (5.13), the one step controllable set $\mathrm{Pre}\{\Omega_x\}$ is written as $\mathrm{Pre}\{\Omega_x\} = \{x \in \Re^n : (\mathbf{A} + \mathbf{KB})x \in \Omega_x\}$. Since the transformation $(A + KB)x$ is linear and satisfies surjection, the mapping $\mathrm{Pre}\{\Omega_x\}$ is a linear mapping. Therefore, when Ω_x is a closed set containing the origin, $\mathrm{Pre}\{\Omega_x\}$ is also closed. As a result, we have both $\mathrm{Pre}\{\Omega_0\}$ and Ω_0 closed. This leads to the fact that the terminal region Ω is a closed set and thus A1 is proved. When the input constraints are given, A2 automatically holds for the linearized model. A3 is also satisfied because we calculate the terminal region as the maximum invariant set of the linearized model at the origin.

To demonstrate A4, we first rewrite A4 as:

$$\begin{aligned} &F(\mathbf{A}\boldsymbol{x}_e + \mathbf{B}\boldsymbol{\tau}) - F(\boldsymbol{x}_e) + \ell(\boldsymbol{x}_e, \boldsymbol{\tau}) \\ &= (\mathbf{A}\boldsymbol{x}_e + \mathbf{B}\boldsymbol{\tau})^\top \mathcal{Q}_f(\mathbf{A}\boldsymbol{x}_e + \mathbf{B}\boldsymbol{\tau}) - \boldsymbol{x}_e^\top \mathcal{Q}_f \boldsymbol{x}_e + \boldsymbol{x}_e^\top Q_f \boldsymbol{x}_e \\ &= ((\mathbf{A} + \mathbf{B}K)\boldsymbol{x}_e)^\top \mathcal{Q}_f((\mathbf{A} + \mathbf{B}K)\boldsymbol{x}_e) - \boldsymbol{x}_e^\top \mathcal{Q}_f \boldsymbol{x}_e + \boldsymbol{x}_e^\top Q_f \boldsymbol{x}_e. \end{aligned} \tag{5.14}$$

By substituting $K = -(\mathbf{B}^\top \mathcal{Q}_f \mathbf{B} + \mathcal{R})^{-1} \mathcal{Q}_f^\top \mathcal{Q}_f \mathbf{A}$, and $\mathcal{Q}_f$ as the solution of the algebraic Riccati equation $\mathcal{Q}_f = \mathbf{A}^\top \mathcal{Q}_f \mathbf{A} + Q_f - \mathbf{A}^\top \mathcal{Q}_f \mathbf{B}(\mathbf{B}^\top \mathcal{Q}_f \mathbf{B} + \mathcal{R})^{-1} \mathbf{B}^\top \mathcal{Q}_f \mathbf{A}$, we have:

$$\begin{aligned} &p(\mathbf{A}\boldsymbol{x}_e + \mathbf{B}\boldsymbol{\tau}) - p(\boldsymbol{x}_e) + q(\boldsymbol{x}_e, \boldsymbol{\tau}) = \\ &\boldsymbol{x}_e^\top \left(\mathbf{A}^\top \left(\mathcal{Q}_f - \mathcal{Q}_f \mathbf{B}(\mathbf{B}^\top \mathcal{Q}_f \mathbf{B} + \mathcal{R})^{-1} \mathbf{B}\mathcal{Q}_f \right) \mathbf{A} + Q_f - \mathcal{Q}_f \right) \boldsymbol{x}_e, \end{aligned} \tag{5.15}$$

where $\mathbf{A}^\top \left(\mathcal{Q}_f - \mathcal{Q}_f \mathbf{B}(\mathbf{B}^\top \mathcal{Q}_f \mathbf{B} + \mathcal{R})^{-1} \mathbf{B}\mathcal{Q}_f \right) \mathbf{A} + Q_f - \mathcal{Q}_f$ is negatively definite. Therefore, assumption A4 is also satisfied.

Given state $\boldsymbol{x}_e(k)$, by solving the optimal control problem (5.11), we have the optimal control input sequence $\boldsymbol{U}^o(k) := [\boldsymbol{\tau}^o(0 \mid k), \boldsymbol{\tau}^o(1 \mid k), \cdots, \boldsymbol{\tau}^o(N_p - 1 \mid k)]$ corresponding to the optimal system trajectory $\boldsymbol{X}_e^o(k) := [\boldsymbol{x}_e(0 \mid k), \boldsymbol{x}_e^o(1 \mid k), \boldsymbol{x}_e^o(2 \mid k), \cdots, \boldsymbol{x}_e^o(N_p \mid k)]$ and the optimal cost function $J_{N_p}^o(k) = J(\boldsymbol{x}_e(k), \boldsymbol{U}^o(k))$, where $\boldsymbol{\tau}^o(i \mid k)$ and $\boldsymbol{x}_e^o(i \mid k)$ are the ith optimal predicted input and states. For the next state $\boldsymbol{x}_e^+ = \boldsymbol{x}_e^o(1 \mid k)$, we can construct a feasible control sequence $\tilde{\boldsymbol{U}}(k+1) = [\boldsymbol{\tau}^o(1 \mid k), \boldsymbol{\tau}^o(2 \mid k), \cdots, \boldsymbol{\tau}^o(N_p - 1 \mid k), K\boldsymbol{x}_e^o(N_p \mid k)]$ and obtain the corresponding state trajectory $\tilde{\boldsymbol{X}}_e(k+1) := [\boldsymbol{x}_e^o(1 \mid k), \boldsymbol{x}_e^o(2 \mid k), \cdots, \boldsymbol{x}_e^o(N_p \mid k), (\mathbf{A} + K\mathbf{B})\boldsymbol{x}_e^o(N_p \mid k)]$ and cost $J_{N_p}(\boldsymbol{x}_e^+) = J\left(\boldsymbol{x}_e^+, \tilde{\boldsymbol{U}}(k+1)\right)$. Since A1-A4 are satisfied, the cost function, which is also the Lyapunov function for the control problem, satisfies the following inequality: Given state $\boldsymbol{x}_e(k)$, by solving the optimal control problem (5.11), we have the optimal control input sequence $\boldsymbol{U}^o(k) := [\boldsymbol{\tau}^o(0 \mid k), \boldsymbol{\tau}^o(1 \mid k), \cdots, \boldsymbol{\tau}^o(N_p - 1 \mid k)]$ corresponding to the optimal system trajectory $\boldsymbol{X}_e^o(k) := [\boldsymbol{x}_e(0 \mid k), \boldsymbol{x}_e^o(1 \mid k), \boldsymbol{x}_e^o(2 \mid k), \cdots, \boldsymbol{x}_e^o(N_p \mid k)]$ and the optimal cost function $J_{N_p}^o(k) = J(\boldsymbol{x}_e(k), \boldsymbol{U}^o(k))$, where $\boldsymbol{\tau}^o(i \mid k)$ and $\boldsymbol{x}_e^o(i \mid k)$ are the ith optimal predicted input and states. For the next state $\boldsymbol{x}_e^+ = \boldsymbol{x}_e^o(1 \mid k)$, we can construct a feasible control sequence $\tilde{\boldsymbol{U}}(k+1) = [\boldsymbol{\tau}^o(1 \mid k), \boldsymbol{\tau}^o(2 \mid k), \cdots, \boldsymbol{\tau}^o(N_p - 1 \mid k), K\boldsymbol{x}_e^o(N_p \mid k)]$ and obtain the corresponding state trajectory $\tilde{\boldsymbol{X}}_e(k+1) := [\boldsymbol{x}_e^o(1 \mid k), \boldsymbol{x}_e^o(2 \mid k), \cdots, \boldsymbol{x}_e^o(N_p \mid k), (\mathbf{A} + K\mathbf{B})\boldsymbol{x}_e^o(N_p \mid k)]$ and cost $J_{N_p}(\boldsymbol{x}_e^+) = J\left(\boldsymbol{x}_e^+, \tilde{\boldsymbol{U}}(k+1)\right)$. Since A1-A4 are satisfied, the cost function, which is also the Lyapunov function for the control problem, satisfies the following inequality:

$$J_{N_p}(\boldsymbol{x}_e^+) \leq J_{N_p}^o(\boldsymbol{x}_e) - \boldsymbol{x}_e^\top Q_f \boldsymbol{x}_e.$$

Then, the closed-loop asymptotic stability of the proposed PL-MPC is demonstrated, which means that as time approaches infinity, the tracking error state converges to zero.

Remark 1. Since the closed-loop stability is established with unknown system uncertainties, a certain degree of robustness is also achieved with the designed controller.

5.5.4 Successive linearizations of PL-MPC

Although the introduction of the DNN learning model compensates for the unmodeled system dynamics, it also increases the computational burden. Therefore, the successive linearizations in the framework of MPC are applied to the nonlinear hybrid model.

The basic idea of successive linearizations [72] is to utilize the whole sequence of the control inputs from a previous time, and calculate a shifted system trajectory for linearizations over all the prediction steps. For numerical simulations, we first discretize the continuous model (5.9) through a zero-order holder as:

$$\boldsymbol{x}_e(k+1) = \boldsymbol{x}_e(k) + \int_{kT}^{(k+1)T} f_e(\boldsymbol{x}_e(t), \boldsymbol{x}_d(t), \boldsymbol{\tau}(t)) \, dt, \tag{5.16}$$

where, for brevity, $f_e(\boldsymbol{x}_e(t), \boldsymbol{x}_d(t), \boldsymbol{\tau}(t))$ stands for the nonlinear dynamics in (5.9).

Step 1: Obtain seed input $\boldsymbol{\tau}^0(i|k)$. The optimal control sequence at the previous step is used as the seed input and $\cdot^0$ represents the seed trajectory. According to the

MPC principle, the first element $\boldsymbol{\tau}(0|k-1)$ of the optimal control sequence $\boldsymbol{U}(k-1)$ is applied to the system and the rest are disregarded. For the linearization at MPC step k, define

$$\boldsymbol{\tau}^0(i \mid k) = \boldsymbol{\tau}(i+1 \mid k-1), \tag{5.17}$$

for $i = 0, 1, \ldots, N_p - 2$ and $\boldsymbol{\tau}^0(N_p - 1 \mid k) = \boldsymbol{\tau}(N_p - 1 \mid k - 1)$.

Step 2: Obtain seed state $\boldsymbol{x}_e^0(i|k)$. Via high-precision ordinary differential equation solvers (such as ODE45 in Matlab), the seed state trajectory $\boldsymbol{x}_e^0(i \mid k)$, $i = 0, 1, \ldots, N_p$ can be generated with the seed input $\boldsymbol{\tau}^0(i|k)$, the current state $\boldsymbol{x}_e^0(0|k) = \boldsymbol{x}_e(k)$ and hybrid-physics model (5.10).

Step 3: Linearize the prediction model, cost function, and constraints on the seed trajectory $(\boldsymbol{x}_e^0(i|k), \boldsymbol{\tau}^0(i|k))$. Define the error between the system trajectory and the seed trajectory as $(\Delta\boldsymbol{x}_e(i \mid k), \Delta\boldsymbol{\tau}(i \mid k))$, and satisfy:

$$\begin{aligned} \boldsymbol{x}_e(i \mid k) &= \boldsymbol{x}_e^0(i \mid k) + \Delta\boldsymbol{x}_e(i \mid k), \\ \boldsymbol{\tau}(i \mid k) &= \boldsymbol{\tau}^0(i \mid k) + \Delta\boldsymbol{\tau}(i \mid k). \end{aligned} \tag{5.18}$$

where,

$$\begin{aligned} \mathbf{J}_{\boldsymbol{x}_e}^c(i \mid k) &= \frac{\partial(f_p + f_l - f_d)}{\partial\boldsymbol{x}_e}\Big|_{(\boldsymbol{x}_e^0(i|k), \boldsymbol{x}_d(i|k), \boldsymbol{\tau}^0(i|k))} \in \Re^{6\times6}, \\ \mathbf{J}_{\boldsymbol{\tau}}^c(i \mid k) &= \frac{\partial(f_p + f_l - f_d)}{\partial\boldsymbol{\tau}}\Big|_{(\boldsymbol{x}_e^0(i|k), \boldsymbol{x}_d(i|k), \boldsymbol{\tau}^0(i|k))} \in \Re^{6\times3}. \end{aligned} \tag{5.19}$$

are the Jacobian state matrix and Jacobian control matrix, respectively. The Jacobian state matrix and the Jacobian control matrix of the DNN model are

$$\begin{aligned} \frac{\partial f_l}{\partial\boldsymbol{x_e}} &\triangleq \begin{bmatrix} \mathbf{0}_{3\times3} & \mathbf{0}_{3\times3} \\ \mathbf{0}_{3\times3} & \frac{\partial f_l}{\partial\boldsymbol{v_e}} \end{bmatrix} \in \Re^{6\times6}, \\ \frac{\partial f_l}{\partial\boldsymbol{\tau}} &\triangleq \begin{bmatrix} \mathbf{0}_{3\times3} \\ \frac{\partial f_l}{\partial\boldsymbol{\tau}} \end{bmatrix} \in \Re^{6\times3}, \end{aligned} \tag{5.20}$$

where

$$\frac{\partial f_l}{\partial\boldsymbol{v_e}} = \frac{\partial f_l}{\partial\boldsymbol{V}} = \tanh'(l_{m-1})\cdots\tanh'(l_0)(1./\sigma_{\boldsymbol{V}})\mathbf{W}_m^T\mathbf{W}_{(m-1)}^T\cdots\mathbf{W}_1^T\sigma_{\boldsymbol{d}} \in \Re^{3\times3},$$

$$\frac{\partial f_l}{\partial\boldsymbol{\tau}} = \tanh'(l_{m-1})\cdots\tanh'(l_0)(1./\sigma_{\boldsymbol{\tau}})\mathbf{W}_m^T\mathbf{W}_{(m-1)}^T\cdots\mathbf{W}_1^T\sigma_{\boldsymbol{d}} \in \Re^{3\times3}.$$

For successive linearizations, the activation function needs to be a derivable function, such as $tanh$. Then, we can get the discrete linearized incremental model

$$\Delta\boldsymbol{x}_e(i+1 \mid k) = \mathbf{J}_{\boldsymbol{x}}^d(i \mid k)\Delta\boldsymbol{x}_e(i \mid k) + \mathbf{J}_{\boldsymbol{\tau}}^d(i \mid k)\Delta\boldsymbol{\tau}(i \mid k). \tag{5.21}$$

According to (5.16), (5.18) and (5.21), the predicted system states $\boldsymbol{x}_e(i|k)$, $i = 1, 2, ..., N_p$, at the time step k, can be expressed as:

$$\boldsymbol{x}_e(1 \mid k) = \boldsymbol{x}_e^0(1 \mid k) + \mathbf{J}_\tau^d(0 \mid k)\Delta\boldsymbol{\tau}(0 \mid k),$$

$$\boldsymbol{x}_e(2 \mid k) = \boldsymbol{x}_e^0(2 \mid k) + \mathbf{J}_x^d(1 \mid k)\mathbf{J}_\tau^d(0 \mid k)\Delta\boldsymbol{\tau}(0 \mid k)$$

$$+ \mathbf{J}_\tau^d(1 \mid k)\Delta\boldsymbol{\tau}(1 \mid k),$$

$$\vdots$$

$$\boldsymbol{x}_e(N_p \mid k) = \boldsymbol{x}_e^0(N_p \mid k) + \mathbf{J}_x^d(N_p - 1 \mid k)\cdots\mathbf{J}_\tau^d(0 \mid k)$$

$$\Delta\boldsymbol{\tau}(0 \mid k) + \cdots + \mathbf{J}_\tau^d(N_p - 1 \mid k)\Delta\boldsymbol{\tau}(N_p - 1 \mid k),$$

(5.22)

with the initial state, define the following vectors:

$$\mathbf{G}(k) = \begin{bmatrix} \mathbf{J}_\tau^d(0 \mid k) & \cdots & \mathbf{0}_{6\times6} \\ \mathbf{J}_x^d(1 \mid k) \cdot \mathbf{J}_\tau^d(0 \mid k) & \cdots & \mathbf{0}_{6\times6} \\ \vdots & \ddots & \vdots \\ \mathbf{J}_x^d(N_p - 1 \mid k) & \cdots & \mathbf{J}_\tau^d(N_p - 1 \mid k) \\ \cdots \mathbf{J}_\tau^d(0 \mid k) & & \end{bmatrix}$$

$$\in \Re^{6N_p \times 3N_p},$$

$$\mathbf{X}_e^0(k) = \begin{bmatrix} \boldsymbol{x}_e^0(1 \mid k)^T & \cdots & \boldsymbol{x}_e^0(N_p \mid k)^T \end{bmatrix} \in \Re^{6N_p \times 1},$$

$$\Delta\mathbf{U} = \begin{bmatrix} \Delta\boldsymbol{\tau}(0 \mid k)^T & \cdots & \Delta\boldsymbol{\tau}(N_p - 1 \mid k)^T \end{bmatrix} \in \Re^{3N_p \times 1}.$$

The predicted system states is then expressed compactly in the following form:

$$\mathbf{X}_e(k) = \mathbf{X}_e^0(k) + \mathbf{G}(k)\Delta\mathbf{U}(k). \tag{5.23}$$

Hence, the original optimization problem (5.11) is approximated as a QP problem

$$\min J(\Delta\mathbf{U}) = 2\tilde{Q}_f\Delta\mathbf{U} + \Delta\mathbf{U}^T\tilde{\mathcal{R}}\Delta\mathbf{U}$$

$$\text{subject to} \quad \boldsymbol{x}_e^0(i \mid k) + \Delta\boldsymbol{x}(i \mid k) \in \Omega_x,$$
$$\boldsymbol{\tau}^0(i \mid k) + \Delta\boldsymbol{\tau}(i \mid k) \in \Omega_\tau, \tag{5.24}$$
$$\boldsymbol{x}_e^0(N_p \mid k) + \Delta\boldsymbol{x}(N_p \mid k) \in \Omega,$$

where,

$$\tilde{\mathcal{R}} = \left(\mathbf{G}(k)^T\bar{Q}_f\mathbf{G}(k) + \bar{\mathcal{R}}\right) \in \Re^{3N_p \times 3N_p},$$

$$\tilde{Q}_f = \mathbf{X}_e^0(k)^T\bar{Q}_f\mathbf{G}(k) \in \Re^{1 \times 3N_p},$$

$$\bar{\mathcal{R}} = \begin{bmatrix} \mathcal{R} & \mathbf{0}_{3\times3} & \cdots & \mathbf{0}_{3\times3} \\ \mathbf{0}_{3\times3} & \mathcal{R} & \cdots & \mathbf{0}_{3\times3} \\ \vdots & \vdots & \ddots & \vdots \\ \mathbf{0}_{3\times3} & \mathbf{0}_{3\times3} & \cdots & \mathcal{R} \end{bmatrix} \in \Re^{3N_p \times 3N_p},$$

TABLE 5.2 CyberShip II Hydrodynamic Parameters [73].

Parameters	Numerical Value	Parameters	Numerical Value
m	23.800	$Y_{\lvert v \rvert v}$	-36.282
I_z	1.760	N_v	0.105
x_g	0.046	$N_{\lvert vv}$	5.044
$X_{\dot{i}}$	-2.000	$Y_{\lvert r \rvert v}$	-0.805
$Y_{\dot{v}}$	-10.000	Y_r	0.108
$Y_{\dot{r}}$	-0.000	$Y_{\lvert v \rvert r}$	-0.845
$N_{\dot{v}}$	-0.000	$Y_{\lvert r \rvert r}$	-3.450
$N_{\dot{r}}$	-1.000	$N_{r \lvert v \rvert}$	0.130
X_u	-0.723	N_r	-1.900
$X_{\lvert u \rvert u}$	-1.327	$N_{\lvert v \rvert r}$	0.080
X_{uuu}	-5.866	$N_{\lvert r \rvert r}$	-0.750
Y_v	-0.861		

$$
\bar{Q}_f =
\begin{bmatrix}
Q_f & \mathbf{0}_{6\times6} & \cdots & \mathbf{0}_{6\times6} \\
\mathbf{0}_{6\times6} & Q_f & \cdots & \mathbf{0}_{6\times6} \\
\vdots & \vdots & \ddots & \vdots \\
\mathbf{0}_{6\times6} & \mathbf{0}_{6\times6} & \cdots & \mathcal{Q}_f
\end{bmatrix}
\in \Re^{6N_p \times 6N_p}.
$$

The optimal incremental control sequence, is provided by solving the Quadratic Programming problem (5.24). The first element of the control sequence is applied to the USV system.

5.6 SIMULATIONS AND DISCUSSIONS

This section describes the numerical simulation results based on the 3 DOF nonlinear USV model of CyberShip II [73].

5.6.1 Simulation settings

The system parameters of CyberShip II are described in Table 5.2 [73]. The simulations in this section are performed on a platform with Intel® Core™ i5-10400f CPU, NVIDIA® GeForce™ GTX 960 and 32GB RAM.

The physics model parameters are identified by using the least-square method (5.5) to process the data of the $20°/20°$ Zigzag trajectory shown in Fig. 5.3. The identified parameters of the physics model are shown in Table 5.3. Because the Zigzag trajectory cannot fully excite the system dynamics, there exist modeling errors between the identified model (5.4) and the true system model (5.2). The DNN-based learning model is therefore proposed to compensate for the modeling errors. The DNN is constructed in the Pytorch environment. The DNN is designed with three hidden layers and each layer with five neurons. The hyperbolic tangent function is used as

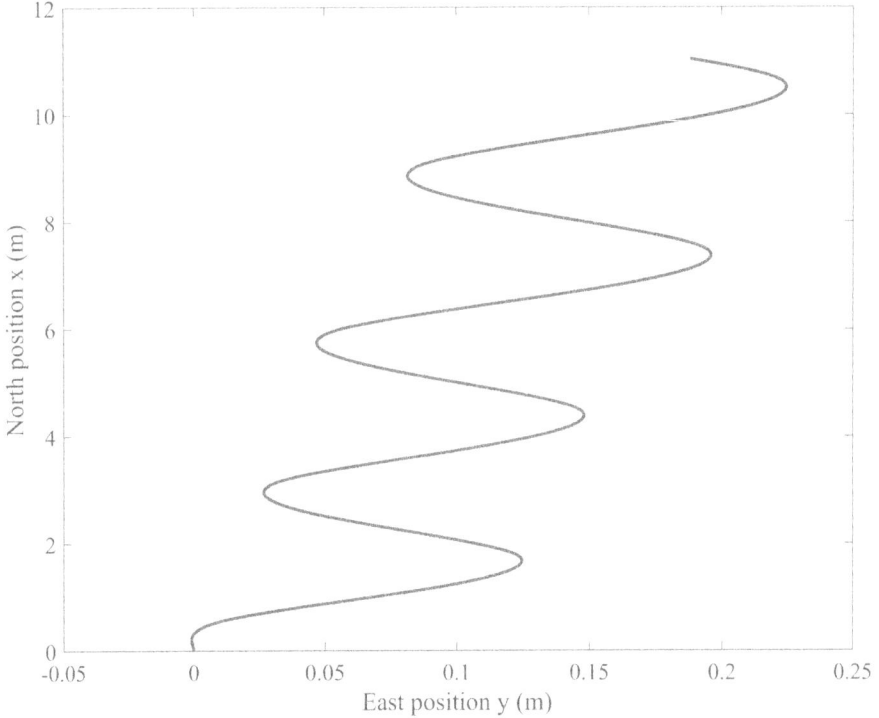

Figure 5.3: The 20°/20° Zigzag trajectory used in the model identification.

the activation function for its differentiable property. Note that both the number of the hidden layers and the number of the layer neurons influence the learning performance of the DNN. In general, too few layers or neurons might result in failures in learning the complex system functions. Too many layers or neurons might lead to vanishing gradients, exploding gradients, or overfitting issues. Therefore, the above DNN structure is designed by considering all these factors and also by referring to the application cases of [74][27]. Inspired by [74] with two hidden layers and 64 neurons in each layer and [27] with two hidden layers and 64/16 neurons in the two layers, we increase the depth of the DNN while reducing the width to improve the learning ability according to [75]. Moreover, ablation experiments have been conducted to evaluate the performance of the designed DNN structure. The DNN training algorithm uses the **Adam** optimizer, with 150000 training times and 0.001 learning rate.

The reference trajectory for the USV to track is:

$$\begin{cases} x(k) = 8\sin(0.02k), \\ y(k) = 8(1 - \cos(0.02k)), \end{cases}$$

for $k = 1, \ldots, 320$. The prediction horizon is set to $N_p = 12$. Note that the proposed PL-MPC and SL-PLMPC controllers are readily extendable to the vertical trajectory tracking tasks if the vertical reference trajectory is available. Disturbance forces are set as: $b_x = 0.4(1 + 0.1r(0,1))\cos(\pi/6)\,\mathrm{N}$, $b_y = 0.4(1 + 0.1r(0,1))\sin(\pi/6)\,\mathrm{N}$, where

TABLE 5.3 CyberShip II system identification parameters and values.

Parameters	Numerical Value	Parameters	Numerical Value
$\hat{a}_{11}$	0.00387	$\hat{a}_{12}$	1.26487
$\hat{a}_{13}$	−0.00564	$\hat{a}_{14}$	−0.00271
$\hat{a}_{21}$	0.02397	$\hat{a}_{22}$	−0.01804
$\hat{a}_{23}$	−0.01899	$\hat{a}_{24}$	−0.79795
$\hat{a}_{25}$	0.05186	$\hat{a}_{26}$	0.05538
$\hat{a}_{31}$	−0.00895	$\hat{a}_{32}$	0.34532
$\hat{a}_{33}$	−2.65462	$\hat{a}_{34}$	0.20365
$\hat{a}_{35}$	−0.04120	$\hat{a}_{36}$	−0.78483

$r(0, 1)$ generates stochastic numbers from a normal distribution with a mean of 0 and a standard deviation of 1.

The system sampling time and linearization time step are both 1 s. The weighting matrices are set as: $Q_f = \text{diag}([80, 80, 100, 50, 50, 100])$, $\mathcal{R} = \text{diag}([10, 10, 10])$, where diag means diagonal matrices. The system constraints due to the physical limitations, control rate saturation and the terminal region are

$$|\tau_{\max}| = [2\,\text{N}, 1.5\,\text{N}, 1.5\,\text{Nm}]^T,$$
$$|\tau_{\max}^r| = [0.5\,\text{N/s}, 0.375\,\text{N/s}, 0.375\,\text{Nm/s}]^T,$$
$$\Omega = \left\{ \mathbf{A} \in \Re^{24 \times 6}, \mathbf{b} \in \Re^{24 \times 1} \mid \mathbf{A}\boldsymbol{x}_e \leq \mathbf{b} \right\},$$

where the terminal region is constrained by $\mathbf{A}$ and $\mathbf{b}$.

In order to fairly compare the effectiveness of various algorithms, simulations are conducted under the same initial state conditions. The simulation is initialized with the states $\boldsymbol{x} = [0.16\,\text{m}, 0\,\text{m}, 0.02\,\text{rad}, 0.05\,\text{m/s}, 0\,\text{m/s}, 0\,\text{rad/s}]^T$.

To demonstrate the effectiveness of the proposed PL-MPC and the efficiency of the proposed Successive Linearization PL-MPC, we next conduct comprehensive comparisons of the proposed three methods with MPC, Linear Quadratic Regulator (LQR) and Koopman-MPC. Since LQR cannot directly consider constraints in the optimization, we limit the control input of LQR by saturation. The implemented MPC algorithm and LQR algorithm use the physical model and the linearized physical model, respectively. Based on [76], we design the implemented Koopman-MPC with Gaussian radial basis functions as the lifting function. In order to accurately learn the USV dynamics, we have carefully tuned the relevant parameters, and set the lifting dimension to 1000. The training data for the Koopman operator and the DNN is from the same trajectory tracking simulation.

5.6.2 Tracking performance

Figs. 5.4 and 5.5 show the comparison results of the control input and velocity state trajectories of the Successive Linearization PL-MPC (SL-PLMPC), PL-MPC, MPC,

LQR and Koopman-MPC. In both Figs. 5.4 and 5.5, it can be seen that, in the lateral motions (Figs. 5.4(b)-(c), Figs. 5.5(b)-(c)), MPC, LQR and Koopman-MPC have much larger fluctuations than that of PL-MPC and SL-PLMPC at the starting stage. The controlled trajectories by MPC and LQR also have slower convergence rates. The larger fluctuations and slower convergence rates are both due to the unidentified modeling dynamics, especially in the sway and yaw motions. Compared to MPC and LQR, Koopman-MPC has higher modeling accuracy and thus better control performance. It is worth noting that around $t = 2\,\mathrm{s}$, the surge force control and surge force control rate of LQR reach saturation. This is caused by the unmodeled dynamics and linearization errors. In addition, PL-MPC and SL-PLMPC exhibit better control performance than that of the Koopman-MPC, indicating that the physical-learning hybrid model has higher modeling accuracy than the Koopman model. Moreover, all the control input trajectories by the four methods are within the system constraints as indicated by the green dotted lines in Fig. 5.4.

The controlled trajectories by both PL-MPC and SL-PLMPC are considerably smoother since the unmodeled system dynamics are compensated with the learning model. Although the SL-PLMPC introduces linearization errors to some extent, it still has comparable control performance with that of PL-MPC. In Fig. 5.4(b), around $t = 2\,\mathrm{s}$, the SL-PLMPC trajectory fluctuates slightly compared with PL-MPC. In fact, this shows that the SL-PLMPC prediction model's accuracy has dropped as a result of linearization. However, since the successive linearization conducts multiple steps of linearizations, the modeling inaccuracy due to linearizations are negligible.

The above observations are further demonstrated in Fig. 5.6 and Table 5.4, where the overall control performance (the cost defined in (11)) is compared. Among these control methods, MPC and LQR using physical models have the highest overall costs, followed by Koopman-MPC. Compared to MPC, LQR has the largest steady-state costs since the linearized physical prediction model is used in LQR. Koopman-MPC utilizes a high-dimensional Koopman operator to learn the USV dynamics, resulting in lower total costs than that of MPC and LQR. The PL-MPC and SL-PLMPC benefit from the physical-learning hybrid model, achieving the even smaller total costs than that of the Koopman-MPC. Figs. 5.7 and 5.8 are further plotted to compare the reference trajectory tracking performance of the five methods. Similar comparison conclusions can be obtained as before. The comparison of tracking errors between PL-MPC, SL-PLMPC, and Koopman-MPC further demonstrates the higher modeling accuracy of the physical-learning hybrid model compared to the Koopman model. It is worth noting that PL-MPC has a larger tracking error overshoot than SL-PLMPC in Fig. 5.7. This is because SL-PLMPC spends larger control efforts at the initial stage.

5.6.3 Computational efficiency comparison

The computational efficiency is important for practical applications. Ideally, the computational time at each time step should be shorter than the sampling time, i.e., $1\,\mathrm{s}$, in our case. The time consumption for MPC, PL-MPC, SL-PLMPC, LQR and Koopman-MPC is shown and compared in Fig. 5.9. Since in PL-MPC, the prediction

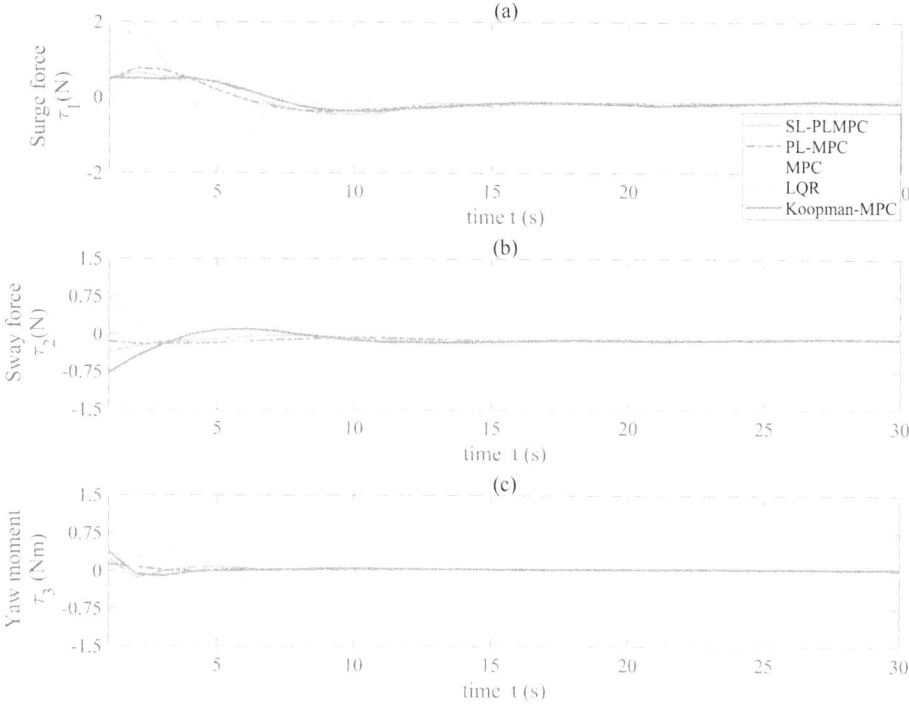

Figure 5.4: Control input trajectories in the first 30 seconds.

model is composed of the physics model and the learning model, PL-MPC has the highest model accuracy and complexity. This will, on the one hand, improve the control performance. On the other hand, it will also lead to the highest computational burden, as shown in Fig. 5.9. At most steps, the computational time of PL-MPC needs more than 60 s for one step. MPC needs relatively shorter computational times compared to PL-MPC. However, this is at the sacrifice of deteriorating the control performance. Moreover, the computational time of MPC at each step is still as high as 40 s which is still not applicable in practice. The LQR method uses a linearized physical model. Therefore, LQR has the smallest computational burden.

The quantitative comparison results on the computational efficiency of MPC, PL-MPC, SL-PLMPC, LQR and Koopman-MPC are further shown in Table 5.5. To approximate the USV dynamics, Koopman-MPC utilizes a high-dimensional Koopman operator. However, this also introduces the potential computational burden, resulting in the lower computational efficiency compared to that of SL-PLMPC. The average computation time for SL-PLMPC is only 0.033 s, with the initialization step and average computation time shorter than the system sampling time of 1 s. This demonstrates the enormous potential of SL-PLMPC in practical applications.

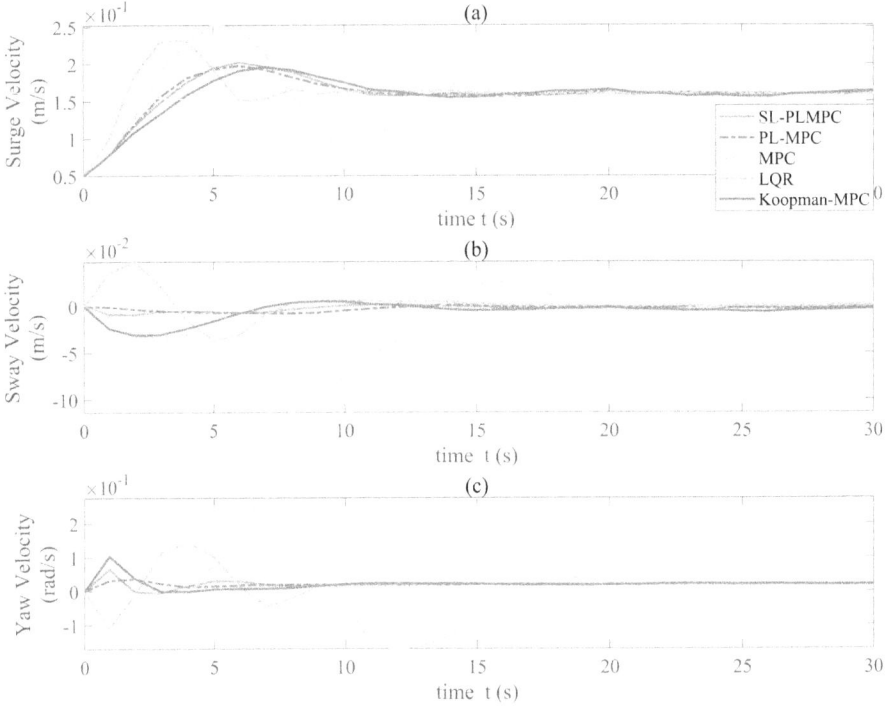

Figure 5.5: Velocity trajectories in the first 30 seconds.

5.6.4 Summary of the comparisons

Based on the above comparisons, it can be concluded that PL-MPC with the hybrid physics-learning model has the best trajectory tracking control performance among the five methods. However, it is also the most computationally heavy, which prohibits its practical applications.

Regarding the computational time, MPC, LQR, and Koopman-MPC outperform PL-MPC. However, LQR and MPC exhibit poor control performance. The Koopman-MPC's control performance is better than that of LQR and MPC, but still worse than that of PL-MPC. SL-PLMPC has the control performance comparable to PL-MPC while achieving the highest computational efficiency. Although PL-MPC has the smallest steady-state cost, SL-PLMPC has excellent performance in computational efficiency, and the steady-state cost of SL-PLMPC is comparable to that of PL-MPC. Furthermore, SL-PLMPC outperforms Koopman-MPC in terms of both computational efficiency and control performance. The above results and analysis directly evaluate the second and the third contributions in Section I. The first contribution of the hybrid physical and learning modeling can be evaluated indirectly since the PL-MPC controller and the successive linearization are designed based on the hybrid model.

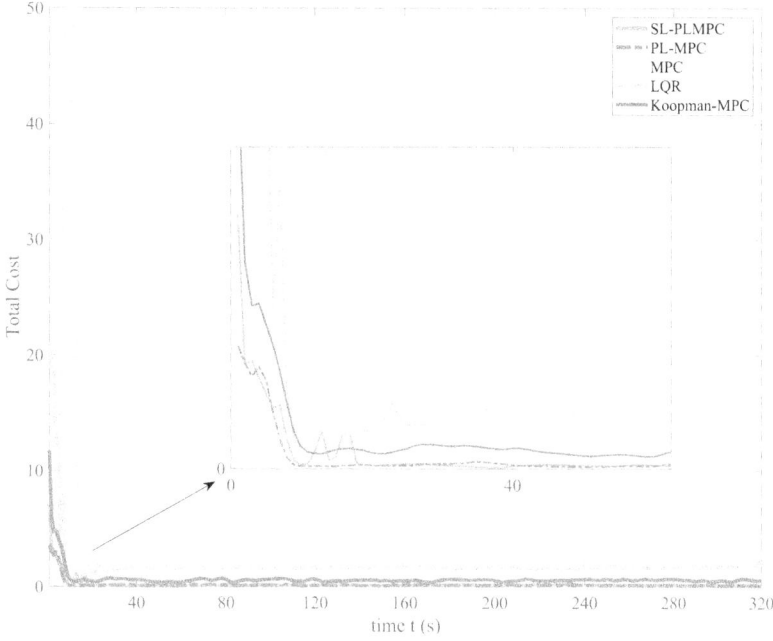

Figure 5.6: Comparison of total costs of the five methods in the whole simulations.

TABLE 5.4   Total cost indicators of the five methods.

	Max	Steady-state	Average
MPC	73.13	0.68	1.88
PL-MPC	3.56	0.09	0.15
SL-PLMPC	7.38	0.10	0.18
LQR	20.27	1.65	1.87
Koopman-MPC	11.76	0.53	0.65

TABLE 5.5   Time consumption of the five methods.

	Max(s)	Min(s)	Average(s)
MPC	44.380	23.528	38.958
PL-MPC	77.204	29.777	60.565
SL-PLMPC	0.837	0.016	0.023
LQR	0.249	0.002	0.004
Koopman-MPC	0.922	0.180	0.218

5.7   CONCLUSIONS

In this chapter, the model free control methods for USVs are discussed, with emphasize on the emerging powerful learning-based approaches. Overall, model free based

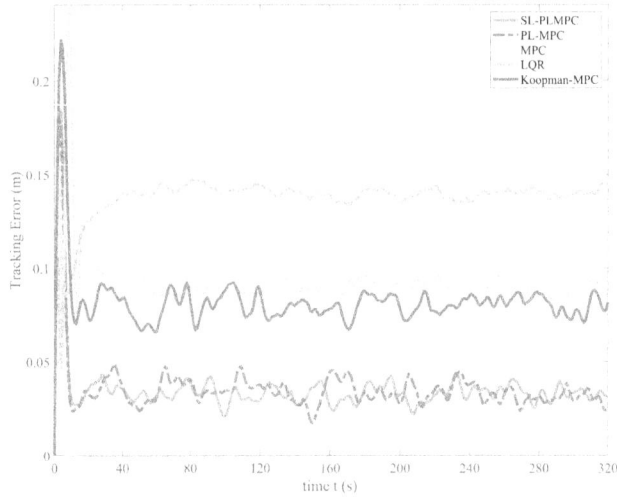

Figure 5.7: Comparison of tracking errors of the five methods in the whole simulations.

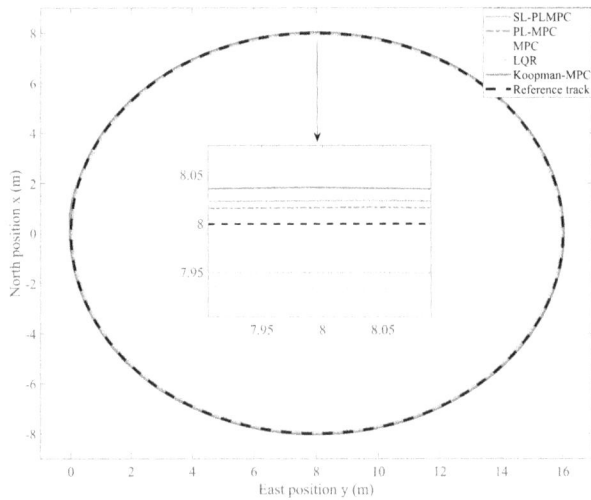

Figure 5.8: Comparison of trajectory tracking of the five methods.

methods do not need to obtain the complex USV hydrodynamics accurately, which is in line with the practical situations. However, model free based methods generally face the "black-box" issue and cannot guarantee the system performance. These issues can be devastating for safety-critical systems or applications.

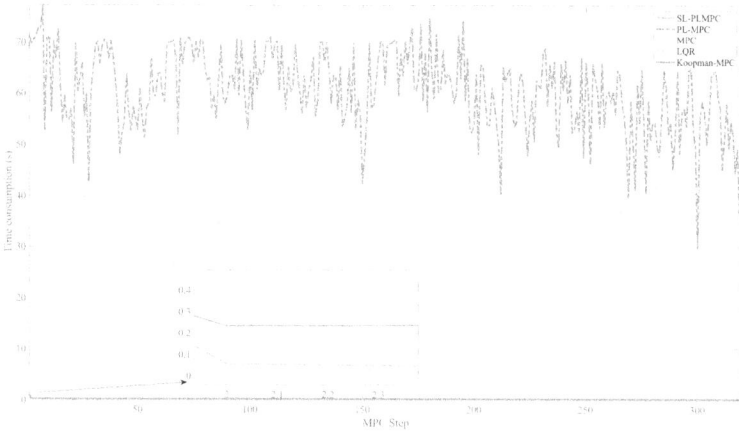

Figure 5.9: Comparison of computing time consumption among various methods in the whole simulations.

Therefore, a new hybrid physics-learning model-based predictive control (PL-MPC) method is proposed that utilize both the USV identified and learned models. From the modeling perspective, we divide the modeling task of a complicated USV system into two parts, i.e., the routinely identified system physical dynamics and the trained unidentifiable system learning dynamics. The learning dynamics are modeled with the Deep Neural Networks (DNN). From the control perspective, we formulate the PL-MPC problem with guaranteed closed-loop stability. Moreover, to achieve a balance between the control performance and the possible computational burden, the successive linearization approach is proposed for the hybrid physics and learning model. Since the modeling, control and verification designs in the simulations are all the same as in the practice, the effectiveness demonstrated via simulation results imply the effectiveness in the practice to a certain extent. The proposed SL-PLMPC not only has the comparable control performance with that of PL-MPC, but also has near real time computational efficiency. SL-PLMPC has the greatest potential in practical applications for USVs. Future work will involve real-world experiments and the exploration of combining offline learning and online adaptive learning to address USV control issues.

Bibliography

[1] N. Minorski, "Directional stability of automatically steered bodies," *Journal of American Society of Naval Engineers*, vol. 42, no. 2, pp. 280–309, 1922.

[2] R. Vilanova and A. Visioli, *PID Control in the Third Millennium.* Springer-Verlag London, UK: Springer, 2012.

[3] F. Yunsheng, S. Xiaojie, W. Guofeng, and G. Chen, "On fuzzy self-adaptive PID control for USV course," in *Proceedings of the 34th Chinese Control Conference,*

Hangzhou, China, 2015, pp. 8472–8478.

[4] J. M. Larrazabal and M. S. Peñas, "Intelligent rudder control of an unmanned surface vessel," *Expert Systems with Applications*, vol. 55, pp. 106–117, 2016.

[5] T. Asfihani, D. K. Arif, F. P. Putra, M. A. Firmansyah *et al.*, "Comparison of LQG and adaptive PID controller for ASV heading control," in *Journal of Physics: Conference Series*, vol. 1218, no. 1, 2019, p. 012058.

[6] X. Xiang, C. Yu, L. Lapierre, J. Zhang, and Q. Zhang, "Survey on fuzzy-logic-based guidance and control of marine surface vehicles and underwater vehicles," *International Journal of Fuzzy Systems*, vol. 20, pp. 572–586, 2018.

[7] X. Cheng, G. Li, R. Skulstad, S. Chen, H. P. Hildre, and H. Zhang, "A neural-network-based sensitivity analysis approach for data-driven modeling of ship motion," *IEEE Journal of Oceanic Engineering*, vol. 45, no. 2, pp. 451–461, 2019.

[8] R. S. Burns, "The use of artificial neural networks for the intelligent optimal control of surface ships," *IEEE Journal of Oceanic Engineering*, vol. 20, no. 1, pp. 65–72, 1995.

[9] Y. Zhang, G. E. Hearn, and P. Sen, "A neural network approach to ship track-keeping control," *IEEE Journal of Oceanic Engineering*, vol. 21, no. 4, pp. 513–527, 1996.

[10] M. Seeger, "Gaussian processes for machine learning," *International Journal of Neural Systems*, vol. 14, no. 02, pp. 69–106, 2004.

[11] R. M. Neal, *Bayesian Learning for Neural Networks*. Springer Science & Business Media, 2012, vol. 118.

[12] Y. Qiao, J. Yin, W. Wang, F. Duarte, J. Yang, and C. Ratti, "Survey of deep learning for autonomous surface vehicles in marine environments," *IEEE Transactions on Intelligent Transportation Systems*, vol. 24, no. 4, pp. 3678–3701, 2023.

[13] D. Ma, X. Chen, W. Ma, H. Zheng, and F. Qu, "Neural network model-based reinforcement learning control for auv 3-D path following," *IEEE Transactions on Intelligent Vehicles*, vol. 9, no. 1, pp. 6208–6220, 2024.

[14] A. B. Martinsen, A. M. Lekkas, S. Gros, J. A. Glomsrud, and T. A. Pedersen, "Reinforcement learning-based tracking control of ASVs in varying operational conditions," *Frontiers in Robotics and AI*, vol. 7, p. 32, 2020.

[15] I. Lenz, R. A. Knepper, and A. Saxena, "Deepmpc: Learning deep latent features for model predictive control." in *Robotics: Science and Systems*, vol. 10. Rome, Italy, 2015.

[16] F. Li, H. Li, and C. Wu, "Gaussian process-based learning model predictive control with application to ASV," *IEEE Transactions on Industrial Electronics*, 2024.

[17] Z. Peng, F. Xia, L. Liu, D. Wang, T. Li, and M. Peng, "Online deep learning control of an autonomous surface vehicle using learned dynamics," *IEEE Transactions on Intelligent Vehicles*, vol. 9, no. 2, pp. 3283 – 3292, 2024.

[18] D. Chwa, "Adaptive neural output feedback tracking control of underactuated ships against uncertainties in kinematics and system matrices," *IEEE Journal of Oceanic Engineering*, vol. 46, no. 3, pp. 720–735, 2020.

[19] E. Kaufmann, L. Bauersfeld, A. Loquercio, M. Müller, V. Koltun, and D. Scaramuzza, "Champion-level drone racing using deep reinforcement learning," *Nature*, vol. 620, no. 7976, pp. 982–987, 2023.

[20] Y. Jin, X. Liu, Y. Shao, H. Wang, and W. Yang, "High-speed quadrupedal locomotion by imitation-relaxation reinforcement learning," *Nature Machine Intelligence*, vol. 4, no. 12, pp. 1198–1208, 2022.

[21] S. S. Øvereng, D. T. Nguyen, and G. Hamre, "Dynamic positioning using deep reinforcement learning," *Ocean Engineering*, vol. 235, p. 109433, 2021.

[22] Y. Zhao, X. Qi, Y. Ma, Z. Li, R. Malekian, and M. A. Sotelo, "Path following optimization for an underactuated ASV using smoothly-convergent deep reinforcement learning," *IEEE Transactions on Intelligent Transportation Systems*, vol. 22, no. 10, pp. 6208–6220, 2020.

[23] Y. Zhao, Y. Ma, and S. Hu, "ASV formation and path-following control via deep reinforcement learning with random braking," *IEEE Transactions on Neural Networks and Learning Systems*, vol. 32, no. 12, pp. 5468–5478, 2021.

[24] J. Woo, C. Yu, and N. Kim, "Deep reinforcement learning-based controller for path following of an unmanned surface vehicle," *Ocean Engineering*, vol. 183, pp. 155–166, 2019.

[25] A. Carron, E. Arcari, M. Wermelinger, L. Hewing, M. Hutter, and M. N. Zeilinger, "Data-driven model predictive control for trajectory tracking with a robotic arm," *IEEE Robotics and Automation Letters*, vol. 4, no. 4, pp. 3758–3765, 2019.

[26] J. Kabzan, L. Hewing, A. Liniger, and M. N. Zeilinger, "Learning-based model predictive control for autonomous racing," *IEEE Robotics and Automation Letters*, vol. 4, no. 4, pp. 3363–3370, 2019.

[27] K. Y. Chee, T. Z. Jiahao, and M. A. Hsieh, "Knode-MPC: A knowledge-based data-driven predictive control framework for aerial robots," *IEEE Robotics and Automation Letters*, vol. 7, no. 2, pp. 2819–2826, 2022.

[28] M. O'Connell, G. Shi, X. Shi, K. Azizzadenesheli, A. Anandkumar, Y. Yue, and S.-J. Chung, "Neural-fly enables rapid learning for agile flight in strong winds," *Science Robotics*, vol. 7, no. 66, p. eabm6597, 2022.

[29] Y. Yan, X. Zhao, S. Yu, and C. Wang, "Barrier function-based adaptive neural network sliding mode control of autonomous surface vehicles," *Ocean Engineering*, vol. 238, p. 109684, 2021.

[30] H. Zheng, J. Li, Z. Tian, C. Liu, and W. Wu, "Hybrid physics-learning model based predictive control for trajectory tracking of unmanned surface vehicles," *IEEE Transactions on Intelligent Transportation Systems*, vol. 25, no. 9, pp. 11522 – 11533, 2024.

[31] Q. Zhang, W. Pan, and V. Reppa, "Model-reference reinforcement learning for collision-free tracking control of autonomous surface vehicles," *IEEE Transactions on Intelligent Transportation Systems*, vol. 23, no. 7, pp. 8770–8781, 2022.

[32] L. Chen, S.-L. Dai, and C. Dong, "Adaptive optimal tracking control of an underactuated surface vessel using Actor–Critic reinforcement learning," *IEEE Transactions on Neural Networks and Learning Systems*, vol. 35, no. 6, pp. 7520 – 7533, 2024.

[33] W. Zhou, Y. Wang, C. K. Ahn, J. Cheng, and C. Chen, "Adaptive fuzzy backstepping-based formation control of unmanned surface vehicles with unknown model nonlinearity and actuator saturation," *IEEE Transactions on Vehicular Technology*, vol. 69, no. 12, pp. 14 749–14 764, 2020.

[34] Y. Weng and N. Wang, "Data-driven robust backstepping control of unmanned surface vehicles," *International Journal of Robust and Nonlinear Control*, vol. 30, no. 9, pp. 3624–3638, 2020.

[35] L. Liu, D. Wang, Z. Peng, and Q.-L. Han, "Distributed path following of multiple under-actuated autonomous surface vehicles based on data-driven neural predictors via integral concurrent learning," *IEEE Transactions on Neural Networks and Learning Systems*, vol. 32, no. 12, pp. 5334–5344, 2021.

[36] R. Skjetne, "The maneuvering problem," Ph.D. dissertation, Norwegian University of Science and Technology, Trondheim, Norway, 2005.

[37] W. B. Klinger, I. R. Bertaska, K. D. von Ellenrieder, and M. R. Dhanak, "Control of an unmanned surface vehicle with uncertain displacement and drag," *IEEE Journal of Oceanic Engineering*, vol. 42, no. 2, pp. 458–476, 2016.

[38] T. I. Fossen, M. Breivik, and R. Skjetne, "Line-of-sight path following of underactuated marine craft," in *Proceedings of the 6th IFAC on Manoeuvring and Control of Marine Craft*, Girona, Spain, 2003, pp. 244–249.

[39] R. Skjetne, T. I. Fossen, and P. V. Kokotović, "Adaptive maneuvering with experiments for a model ship in a marine control laboratory," *Automatica*, vol. 41, no. 2, pp. 289–298, 2005.

[40] S. J. Ohrem, H. B. Amundsen, W. Caharija, and C. Holden, "Robust adaptive backstepping dp control of rovs," *Control Engineering Practice*, vol. 127, p. 105282, 2022.

[41] A. Gonzalez-Garcia and H. Castañeda, "Guidance and control based on adaptive sliding mode strategy for a ASV subject to uncertainties," *IEEE Journal of Oceanic Engineering*, vol. 46, no. 4, pp. 1144–1154, 2021.

[42] N. Wang, Y. Gao, H. Zhao, and C. K. Ahn, "Reinforcement learning-based optimal tracking control of an unknown unmanned surface vehicle," *IEEE Transactions on Neural Networks and Learning Systems*, vol. 32, no. 7, pp. 3034–3045, 2020.

[43] J. Rodriguez, H. Castañeda, A. Gonzalez-Garcia, and J. Gordillo, "Finite-time control for an unmanned surface vehicle based on adaptive sliding mode strategy," *Ocean Engineering*, vol. 254, p. 111255, 2022.

[44] X. Jiang, G. Xia, Z. Feng, and Z.-G. Wu, "Nonfragile formation seeking of unmanned surface vehicles: A sliding mode control approach," *IEEE Transactions on Network Science and Engineering*, vol. 9, no. 2, pp. 431–444, 2021

[45] H. Abrougui, S. Nejim, S. Hachicha, C. Zaoui, and H. Dallagi, "Modeling, parameter identification, guidance and control of an unmanned surface vehicle with experimental results," *Ocean Engineering*, vol. 241, p. 110038, 2021.

[46] W. Naeem, T. Xu, R. Sutton, and A. Tiano, "The design of a navigation, guidance, and control system for an unmanned surface vehicle for environmental monitoring," *Proceedings of the Institution of Mechanical Engineers, Part M: Journal of Engineering for the Maritime Environment*, vol. 222, no. 2, pp. 67–79, 2008.

[47] G. Rigatos, "A nonlinear optimal control approach for unmanned surface vessels," *Marine Systems & Ocean Technology*, vol. 18, pp. 1–22, 2023.

[48] Y. Zhang, S. Li, and J. Weng, "Learning and near-optimal control of underactuated surface vessels with periodic disturbances," *IEEE Transactions on Cybernetics*, vol. 52, no. 8, pp. 7453–7463, 2021.

[49] H. Zheng, R. R. Negenborn, and G. Lodewijks, "Predictive path following with arrival time awareness for waterborne AGVs," *Transportation Research Part C: Emerging Technologies*, vol. 70, pp. 214–237, 2016.

[50] H. Zheng, "Coordination of Waterborne AGVs," Ph.D. dissertation, Delft University of Technology, Delft, The Netherlands, 2016.

[51] ——, "Robust distributed predictive control of waterborne AGVs—A cooperative and cost-effective approach," *IEEE Transactions on Cybernetics*, vol. 48, no. 8, pp. 2449–2461, 2018.

[52] Z. Tian, H. Zheng, and W. Xu, "Path following of autonomous surface vehicles with line-of-sight and nonlinear model predictive control," in *Proceedings of the 6th International Conference on Transportation Information and Safety*, Wuhan, China, 2021, pp. 1269–1274.

[53] A. Tsolakis, R. R. Negenborn, V. Reppa, and L. Ferranti, "Model predictive trajectory optimization and control for autonomous surface vessels considering traffic rules," *IEEE Transactions on Intelligent Transportation Systems*, 2024.

[54] Z. Peng, B. Zhang, O. Sun, D. Wang, M. Han, L. Liu, and H. Wang, "Finite-set model predictive speed and heading control of autonomous surface vehicles with unmeasured states," in *Proceedings of Tenth International Conference on Intelligent Control and Information Processing*, Marrakesh, Morocco, 2019, pp. 158–163.

[55] H. Wei and Y. Shi, "MPC-based motion planning and control enables smarter and safer autonomous marine vehicles: Perspectives and a tutorial survey," *IEEE/CAA Journal of Automatica Sinica*, vol. 10, no. 1, pp. 8–24, 2023.

[56] X. Wang, Z. Zou, T. Li, and W. Luo, "Path following control of underactuated ships based on nonswitch analytic model predictive control," *Journal of Control Theory and Applications*, vol. 8, no. 4, pp. 429–434, 2010.

[57] N. Wang, Y. Gao, Y. Liu, and K. Li, "Self-learning-based optimal tracking control of an unmanned surface vehicle with pose and velocity constraints," *International Journal of Robust and Nonlinear Control*, vol. 32, no. 5, pp. 2950–2968, 2022.

[58] A. Gonzalez-Garcia, I. Collado-Gonzalez, R. Cuan-Urquizo, C. Sotelo, D. Sotelo, and H. Castañeda, "Path-following and lidar-based obstacle avoidance via nmpc for an autonomous surface vehicle," *Ocean Engineering*, vol. 266, p. 112900, 2022.

[59] Z. Peng, B. Zhang, O. Sun, D. Wang, M. Han, L. Liu, and H. Wang, "Finite-set model predictive speed and heading control of autonomous surface vehicles with unmeasured states," in *Proceedings of Tenth International Conference on Intelligent Control and Information Processing*, Marrakesh, Morocco, 2019, pp. 158–163.

[60] L. Song, C. Xu, L. Hao, J. Yao, and R. Guo, "Research on PID parameter tuning and optimization based on SAC-auto for ASV path following," *Journal of Marine Science and Engineering*, vol. 10, no. 12, p. 1847, 2022.

[61] W. Song, Y. Li, and S. Tong, "Fuzzy finite-time H_∞ hybrid-triggered dynamic positioning control of nonlinear unmanned marine vehicles under cyber-attacks," *IEEE Transactions on Intelligent Vehicles*, vol. 9, no. 1, pp. 970 – 980, 2024.

[62] W. Yuan and X. Rui, "Deep reinforcement learning-based controller for dynamic positioning of an unmanned surface vehicle," *Computers and Electrical Engineering*, vol. 110, p. 108858, 2023.

[63] S. Sivaraj, S. Rajendran, and L. P. Prasad, "Data driven control based on deep q-network algorithm for heading control and path following of a ship in calm water and waves," *Ocean Engineering*, vol. 259, p. 111802, 2022.

[64] G. Zhang, S. Chu, W. Zhang, and C. Liu, "Adaptive neural fault-tolerant control for ASV with the output-based triggering approach," *IEEE Transactions on Vehicular Technology*, vol. 71, no. 7, pp. 6948–6957, 2022.

[65] P.-F. Xu, C.-B. Han, H.-X. Cheng, C. Cheng, and T. Ge, "A physics-informed neural network for the prediction of unmanned surface vehicle dynamics," *Journal of Marine Science and Engineering*, vol. 10, no. 2, p. 148, 2022.

[66] G. X. Wu, Y. Ding, T. Tahsin, and A. Incecik, "Adaptive neural network and extended state observer-based non-singular terminal sliding mode tracking control for an underactuated ASV with unknown uncertainties," *Applied Ocean Research*, vol. 135, p. 103560, 2023.

[67] M. A. Nielsen, *Neural Networks And Deep Learning*. San Francisco, CA, USA, 2015, vol. 25.

[68] T. I. Fossen, *Handbook of Marine Craft Hydrodynamics and Motion Control (Second Edition)*. Hoboken, NJ, USA: Wiley, 2021.

[69] Z. Lu, J. Li, P. Wu, H. Zheng, and Y. Huang, "Hydrodynamic model identification of an unmanned surface vessel with experiment data," *in Proceedings of 2022 6th CAA International Conference on Vehicular Control and Intelligence (CVCI), Nanjing, China, 2022, pp. 1–6*.

[70] C. Shen, "Motion control of autonomous underwater vehicles using advanced model predictive control strategy," Ph.D. dissertation, University of Victoria, Victoria, Canada, 2018.

[71] H. Chen and F. Allgöwer, "A quasi-infinite horizon nonlinear model predictive control scheme with guaranteed stability," *Automatica*, vol. 34, no. 10, pp. 1205–1217, 1998.

[72] H. Zheng, R. R. Negenborn, and G. Lodewijks, "Fast ADMM for distributed model predictive control of cooperative waterborne AGVs," *IEEE Transactions on Control Systems Technology*, vol. 25, no. 4, pp. 1406–1413, 2017.

[73] R. Skjetne, Ø. N. Smogeli, and T. I. Fossen, "A nonlinear ship manoeuvering model: Identification and adaptive control with experiments for a model ship," *Modeling, Identification and Control*, vol. 25, no. 1, pp. 3–27, 2004.

[74] N. A. Spielberg, M. Brown, and J. C. Gerdes, "Neural network model predictive motion control applied to automated driving with unknown friction," *IEEE Transactions on Control Systems Technology*, vol. 30, no. 5, pp. 1934–1945, 2022.

[75] Z. Lu, H. Pu, F. Wang, Z. Hu, and L. Wang, "The expressive power of neural networks: A view from the width," *in Proceedings of the 31st International Conference on Neural Information Processing Systems (NeurIPS), California, USA, 2017, pp. 6231–6239.*

[76] M. Korda and I. Mezić, "Linear predictors for nonlinear dynamical systems: Koopman operator meets model predictive control," *Automatica*, vol. 93, pp. 149–160, 2018.

II

Practices in the Field

CHAPTER 6

USV Systematic Architecture

6.1 INTRODUCTION

BESIDES the key research methodologies on NGC, the systematic design practices are also essential for USVs to move from the pure research concept to conducting field operations, and finally to the mature commercial products. This chapter concludes the experiences that the authors have gained from the practical designs and tests of USVs. Particularly, the USV core hardware components, the generic USV software architecture, and the USV test procedures will be elaborated.

6.2 USV HARDWARE ESSENTIAL ELEMENTS

USVs share five fundamental system components: (1) The ship hull, (2) The power and propulsion system, (3) The perception system, (4) The communication and remote-control system, and (5) The computing system (Fig. 6.1).

- *The ship hull*: Depending on specific applications, the USV hull can be mono-hull [24], catamaran [23], trimaran [25] or even with new concept designs [26]. In general, multi-hull structures have better stability in terms of resisting the environmental disturbances at the expense of higher assembly complexity. Dimensions of USVs vary from meters to tens of meters in length. Usually, the dimensions of USVs for scientific uses are small and for civilian uses are large. The world first autonomous commercial ship, Yara Birkland [27], has a length of 80 meters and tonnage of 3200 dead-weight tons. The dimension of USV hulls should be designed to satisfy the functional requirements on capacity, deck area, buoyancy, etc. To carry containers, passengers, or scientific and military payloads. The materials of the USV hull are mostly carbon fiber, aluminum alloy or titanium alloy due to their strength and light weight. Inflatable USV hulls [28] are also seen due to the low cost.

- *The power and propulsion system*: Different power supply systems can be used for different types of USVs. For instance, in ENDURUNS Project, fuel cells are

DOI: 10.1201/9781003660590-6

adopted for USVs [29]. With the improvement of the battery energy density, it becomes popular to use battery packs as the power source for USVs [30], such as Yara Birkland [27]. The relevant motors and propellers are chosen as the propulsion system. In addition, other propulsion plants, such as waterjet propulsion [31] and pod propulsion [32], are also used for special purposes. Considering both the efficiency and low-carbon purposes, the hybrid power system is also commonly seen, e.g., Changhang Cargo 001 with a diesel main engine + shaft motor + LNG-fueled generator set + lithium battery [33].

- *The perception system*: The perception system of a USV consists of various sensors to provide necessary information on the USV itself and the navigation environment. With the perception system, the status of the ship hull, ship power and propulsion system, cargo, ship motion, and environment (waves, currents, and winds, etc.) can be monitored and evaluated, i.e., maintaining situation awareness [34, 35]. Specifically, the high-precision motion perception can be realized with the differential GNSS (GPS, Beidou, etc.), and INS. The obstacle perception can be realized with the marine radar with Automatic Radar Plotting Aid (ARPA), 3D LiDAR, millimeter-wave radar, visible light and infrared camera, AIS, etc. The navigation environment and navigation mark can be obtained with meteorology and hydrology sensors and the ENC. For the autonomous navigation, several mature intelligent perception systems have been seen in the market, e.g., the ABB Ability Marine Pilot Vision [1], Maersk Group and Sea Machines perception system [36], Rolls Royce intelligent sensing systems [37], and Kongsberg intelligent situation awareness system [38].

- *The communication and remote-control system*: The ship communication system is to realize the data exchange between onboard devices, between ships and ships, and between ships and the shore. Different from terrestrial vehicles devices of which are generally connected with a CAN network, the ship data exchange onboard has no standard communication protocol. Choices among TCP/IP, UDP/IP, CAN, and R485/232 can be made as the interacting network. Moreover, a USV can receive messages or commands from a control center via radio, Wi-Fi, cellular network, or satellites [39]. Reliable remote-control is essential when there exists any malfunction of the USV autonomous navigation system. The remote-control can either be line-of-sight and non-line-of-sight. It is in general more challenging to remotely control the USV in the non-line-of-sight mode since the communication bandwidth and delay requirements would be stringent. The automated 135-meter barge, ZONGA, was remotely controlled from the Seafar control center in 2020 [5]. The autonomous passenger ferry prototype, milliAmpere [40], can be highly maneuverable either by using a joystick steering from a remote-control unit or by the autonomous system.

- *The computing system*: Experienced captains can integrate all of the available information to make proper decisions such as routing, acceleration or deceleration, so as to minimize the fuel consumption and sailing time while still guaranteeing safety. For USVs, all these perception, decision-making, and operation processing need to be handled by the computing units with relevant algorithms. With the evolution of low power and high-performance computing chips, it is possible to apply the increasingly complex algorithms [41] to upgrade the computing capabilities of USVs. Moreover, the cloud computing technology can transfer the mass calculation to the data processing center on the cloud [6], which can largely alleviate the burden of the onboard computing.

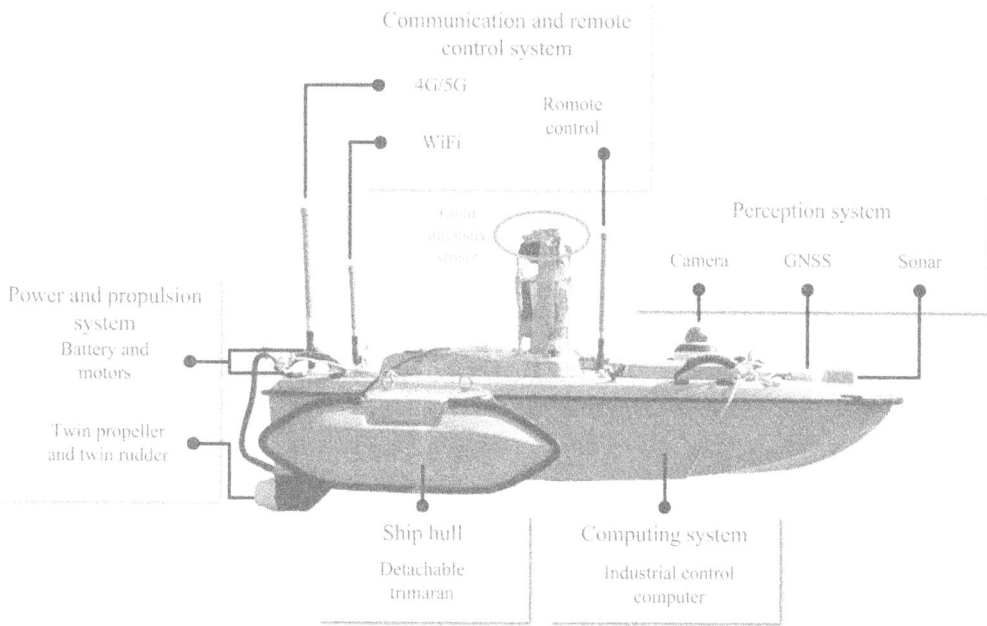

Figure 6.1: Five essential elements of a trimaran USV.

6.3 USV SOFTWARE ARCHITECTURE

Compared with the terrestrial or aerial unmanned vehicles, the generic software specially designed for USVs is less frequently seen. However, the software for other types of unmanned vehicles could also be applied to USVs with moderate modifications.

Especially designed for autonomous marine vehicles, MOOS-IvP [42] provides a set of open source C++ modules, and is developed by the group of researchers from MIT. The core modules of MOOS-Ivp contain the core autonomy, helm behaviors, mission control, etc. Besides, the MSS toolbox [43] is developed as a Matlab and Simulink library for the marine systems and control design. MSS is often used for the proof of principle on small-scale lab USV platforms [44]. Most of the other USV software systems are commercial products and are not open-source, e.g., Transas navigation simulation system [45], Kongsberg autonomous system [8], and Avikus HiNas navigation system [9]. Notably, K-MATE is an autonomous system for executing mission plans developed by Kongsberg, and has been applied in SEA-KIT USV

TABLE 6.1   Major software in the literature/market for USVs.

Software	Organization	Application	Main features
MOOS-IvP [16]	MIT	USVs or AUVs	Software suite for robot autonomy
MSS [43]	NTNU	Cybership [44]	Matlab Toolbox
CaraCas [17]	NASA	USV Swarm II [48]	Autonomy architecture (software infrastructure, core executive functions, technology modules)
K-MATE [47]	Kongsberg	SEA-KIT USV, and YARA Birkeland	Common autonomous control engine
USVs-Sim [18]	Shanghai University	USV navigation generation and control testing	General simulation platform including water, environment, infrastructure, vessel, sensors, data generation and analysis module

and YARA Birkeland [47]. CaraCas is a tightly integrated system to provide robust control for USVs [10, 17].

It should be noted that open-source software for other types of unmanned systems is mostly more powerful since the number of developers and users are larger. Well-known middleware for robotic platforms include Pixhawk [12], Autopilot [13], Robot Operation System (ROS) [14], and Ardupilot [15]. These software systems have been widely utilized for unmanned aerial vehicles and unmanned cars. Both the reliability and capabilities have been widely approved. These software systems can also be applied to USVs with certain adaptations. Briefly, the software differences between USVs and other vehicles mainly exist on the following three aspects: (1) The core sensors for the USV are different, e.g., the marine radar and AIS are the typical sensors for sensing the environment in USVs; (2) It is common to integrate the ENC in the USV software for the maritime navigation; (3) The navigation rules (e.g., collision avoidance, berthing, etc.) for USVs are equivocal. The aforementioned relevant software systems of USVs are listed in Table 6.1.

A typical architecture of USV software is shown in Fig. 6.2. The modular programming idea can improve the efficiency of the software development and debugging. There are three modules composing of the USV software, which are the navigation module, system monitoring module, communication and interface module. For the navigation module, the data processing unit is to parse, filter and analyze different

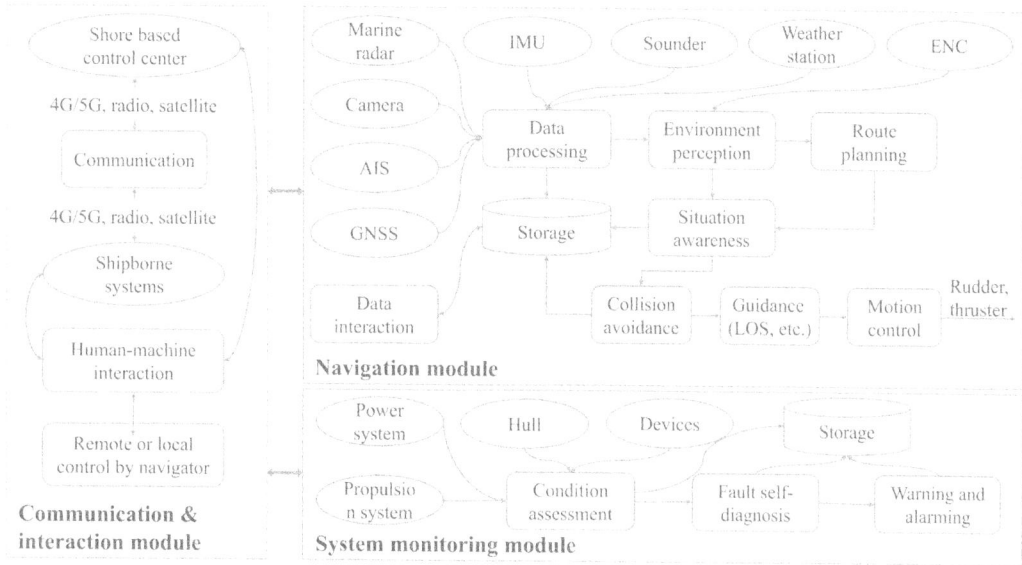

Figure 6.2: Generic architecture of the USV software.

data from the multi-type sensors including GNSS, AIS, marine radar, sounder, etc. The environment perception unit is to sense the navigation scenario, obtain the information of obstacles, and compute the feasible navigation area. Then, in the situation awareness unit, USV navigation status (safety, efficiency, etc.) is evaluated and described. The collision avoidance unit is to deal with the path and motion planning to avoid the dynamic and static obstacles considering the navigation rules. The motion control unit, e.g., the course keeping autopilot, is to follow or track the path or trajectory generated by the guidance unit. All the necessary data are stored with the storage unit. For the system monitoring module, the condition assessment unit can evaluate the status of power system, propulsion system, hull, and other devices, which is used to diagnose the system fault and failure and provide the warning and alarming function. For the communication and interaction module, shore-based control center and shipborne systems are connected with the communication unit (4G/5G, radio or satellite), and the Human-Machine Interaction (HMI) unit is developed to assist the navigator. Moreover, the remote or local control unit can manipulate the USV with the information of the navigation module.

Based on the above software architecture, a specific USV software based on the Qt platform and ENC files is developed to verify the effectiveness of the decision-making and motion control methods. The software interface is shown in Fig. 6.3.

6.4 USV FIELD TEST EXPERIENCES

As the progress of USVs, there is an increasing request to test the performance of the relevant technologies and the reliability of the overall system [20]. The test sites for USVs cover from indoor pools, outdoor pools, lakes, inland rivers to oceans with increasing challenges.

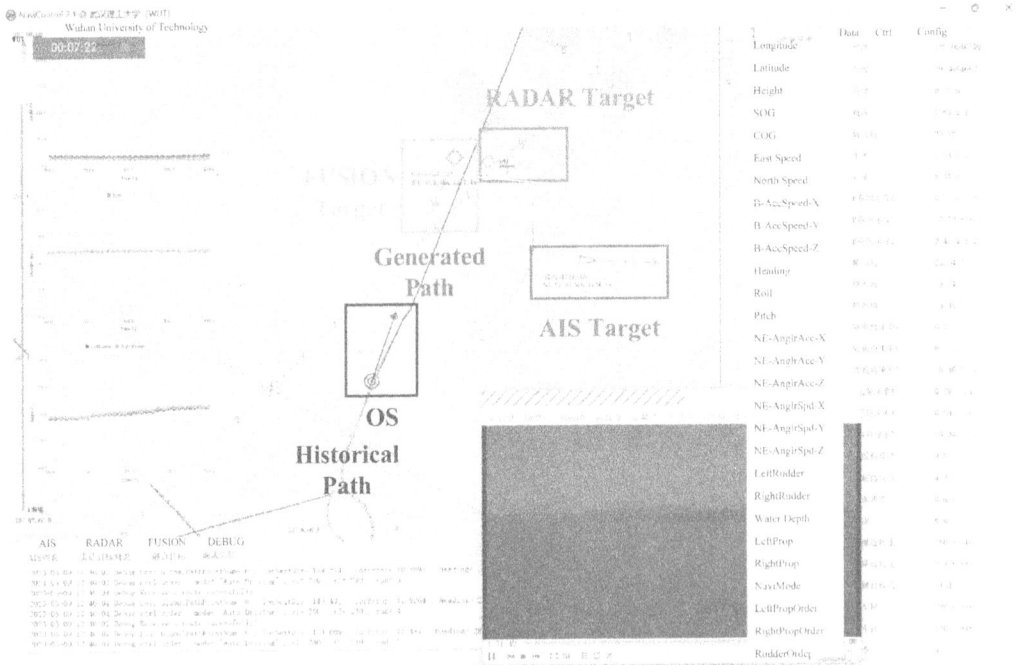

Figure 6.3: A typical USV software interface developed based on Qt platform and ENC files [49].

There are different purposes at different test sites and stages. Fig. 6.4 shows the features of different field test sites for USVs and the relations between them. During the development of the USV methods, technologies and systems, these test scenarios with different scale ships are mutually connected. For instance, in the NOVIMAR project [50], a model ship formation in pool and full scale ship formation in inland/short sea are both tested for the validation of the "vessel train" concept and relevant technologies.

Following the test procedures in Fig. 6.4, a series of USV tests have been tested in East China Sea. This USV is a trimaran ship equipped with the maritime radar, AIS, IMU, ENC, etc. The test scenarios and experiments are shown in Fig. 6.5. During the real experiments, it is found that multiple scenarios need to be set with a real own ship (OS) and target ships (TSs). Therefore, this process might be costly and increase the collision risk. Therefore, the virtual-reality integration test method [51] is also commonly used for USV tests. In such cases, the collision avoidance scenario could be generated with the real OS and the virtual TS. The navigation states of TS can be simulated with historical data of TS or motion model combined with environmental disturbances. This kind of virtual-reality integration test can largely improve the testing efficiency with low costs and risks.

Overall, the procedure for the field tests of USVs can be summarized as three fundamental stages: experimentation, validation, and evaluation [20]. Broadly, the field test items of USVs involve the mooring test, maneuvering test, sea-keeping test, autonomous navigation test, etc. [22]. Most tests focus on the autonomous

| Filed test sites | Features | Relations |

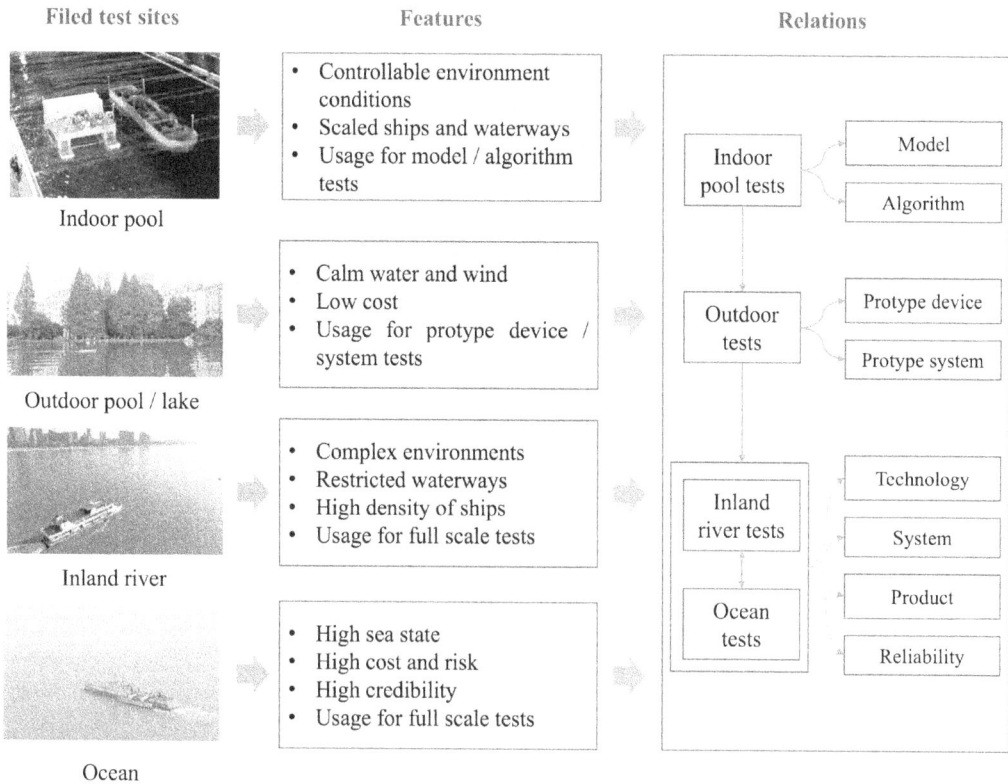

Figure 6.4: Test sites from indoor pools to oceans.

maneuvering capabilities consisting of the perceiving and recognition subsystem, decision-making subsystem, motion control subsystem and remote-control subsystem. These tests cover the functions and reliability of each subsystem, and also the integrated performance of different subsystems.

Based on the authors' field test experiences, the complete USV test considerations are concluded in Table 6.2.

Conducting real-sea trials for collision avoidance systems of USVs entails formidable challenges, given the generally low intelligence level of current ships. Fundamental prerequisites for these trials include digital signal control for propulsion and steering systems, feedback mechanisms, and the integration of diverse sensor data. In our future work, it is planned to explore collision avoidance strategies that involve changes in both speed and heading to better comply with collision avoidance regulations. Additionally, we aim to select controllers that more accurately reflect ship avoidance behavior to minimize the discrepancy between planned and actual paths. Furthermore, we intend to address the challenge of multi-ship encounters by employing strategies such as fuzzy logic to further delineate the own ship avoidance responsibilities within complex encounter scenarios.

TABLE 6.2   USV test considerations based on the authors' experiences.

Subsystem	Functionality testing	Performance testing	Reliability testing	Intelligence testing
Perceiving and recognition subsystem	1. Environment sensing 2. Object detection 3. Situation awareness	1. Task planning 2. Collision avoidance 3. Energy saving 4. Autonomy 5. Safety	1. Target recognition 2. False alarm 3. Fault-tolerance 4. Robustness 5. Resilience 6. Instantaneity	1. Human-like testing 2. Turning test 3. Fault self-diagnosis 4. Self-learning 5. Early warning
Decision-making subsystem	1. Route planning 2. Motion prediction 3. Planned trajectory 4. Optimized speed			
Motion control subsystem	1. Path following 2. Trajectory tracking 3. Posture control			
Remote-control subsystem	1. Remote monitoring 2. Control switching 3. Remote-control			

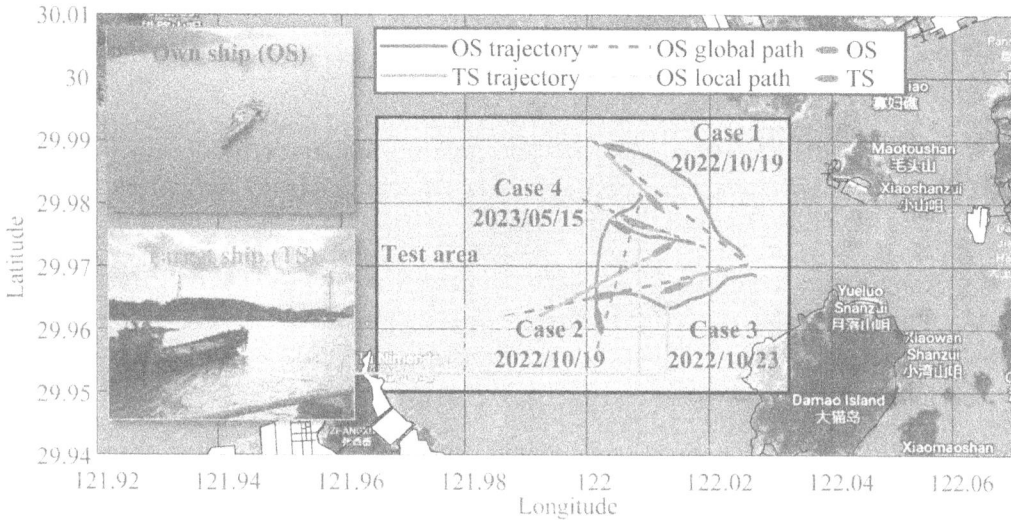

Figure 6.5: Test scenarios and experiments in East China Sea [49].

6.5 CONCLUSIONS

In this chapter, we propose a system framework including hardware and software for USVs, and the experiments of USVs were conducted in real coastal waters. It is a systematic project to build a USV of conducting different tasks, here we mainly focused on the common methods and technologies of navigation safety and intelligence. More details can be found in the literature, e.g., [49, 33, 4, 52].

Bibliography

[1] ABB, "ABB Abilit Marine Pilot Vision," https://new.abb.com/marine/systems-and-solutions/digital/abb-ability-marine-pilot/abb-ability-marine-pilot-vision, 2017, [Online; 'accessed 7-July-2024].

[2] Maersk, "Maersk selects Sea Machines for world's first AI-powered situational awareness system aboard a container ship," https://www.marinelink.com/news/situational-technology436795, 2018, [Online; 'accessed 7-July-2024].

[3] Rolls-Royce, "Intelligent awareness: Enhancing safety through intelligent data fusion," https://www.ferryshippingsummit.com/wp-content/uploads/2018/03/Rolls-Royce.pdf, 2018, [Online; 'accessed 7-July-2024].

[4] Kongsberg, "Intelligent awareness," https://www.kongsberg.com/maritime/the-full-picture-magazine/2018/6/intelligent-awareness/, 2019, [Online; 'accessed 7-July-2024].

[5] Seafar, "Shore supported navigation for 135m barge," https://seafar.eu/shore-supported-navigation-for-135m-barge/, 2020, [Online; 'accessed 7-July-2024].

[6] J. Guo and H. Guo, "Real-time risk detection method and protection strategy for intelligent ship network security based on cloud computing," *Symmetry*, vol. 15, no. 5, p. 988, 2023.

[7] Transas, "WECDIS equipment and marine simulators for naval training," `https://www.naval-technology.com/contractors/software/transas/`, 2024, [Online; accessed '7-July-2024].

[8] Kongsberg, "Autonomy is here – powered by KONGSBERG," `https://www.kongsberg.com/maritime/about-us/news-and-media/our-stories/autonomy-is-here--powered-by-kongsberg/`, 2024, [Online; 'accessed 9-July-2024].

[9] Avikus, "HiNAS navigation," `https://avikus.ai/hinas-navigation`, 2024, [Online; accessed '9-July-2024].

[10] T. Huntsberger, H. Aghazarian, D. Gaines, M. Garrett, and G. Sibley, "Autonomous operation of unmanned surface vehicles (USVs)," in *Proc. IEEE ICRA Workshop on Robots in Challenging and Hazardous Environments*, Citeseer, 2007.

[11] M. T. Wolf, A. Rahmani, J.-P. de la Croix, G. Woodward, J. V. Hook, D. Brown, S. Schaffer, C. Lim, P. Bailey, S. Tepsuporn, M. Pomerantz, V. Nguyen, C. Sorice, and M. Sandoval, "CARACaS multi-agent maritime autonomy for unmanned surface vehicles in the swarm II harbor patrol demonstration," in *Proceedings of the SPIE*, p. 10195, 2017.

[12] L. Meier, P. Tanskanen, F. Fraundorfer, and M. Pollefeys, "Pixhawk: A system for autonomous flight using onboard computer vision," in *2011 IEEE International Conference on Robotics and Automation*, IEEE, pp. 2992–2997, 2011.

[13] R. L. Ribler, J. S. Vetter, H. Simitci, and D. A. Reed, "Autopilot: Adaptive control of distributed applications," in *Proceedings. The Seventh International Symposium on High Performance Distributed Computing (Cat. No. 98TB100244)*, IEEE, pp. 172–179, 1998.

[14] S. Macenski, T. Foote, B. Gerkey, C. Lalancette, and W. Woodall, "Robot operating system 2: Design, architecture, and uses in the wild," *Science Robotics*, vol. 7, no. 66, p. eabm6074, 2022.

[15] S. Baldi, D. Sun, X. Xia, G. Zhou, and D. Liu, "Ardupilot-based adaptive autopilot: Architecture and software-in-the-loop experiments," *IEEE Transactions on Aerospace and Electronic Systems*, vol. 58, no. 5, pp. 4473–4485, 2022.

[16] N. C. Evans, "A practical search with Voronoi distributed autonomous marine swarms," Ph.D. dissertation, Massachusetts Institute of Technology, 2022.

[17] M. T. Wolf, A. Rahmani, J.-P. de la Croix, G. Woodward, J. Vander Hook, D. Brown, S. Schaffer, C. Lim, P. Bailey, S. Tepsuporn, *et al.*, "Caracas multi-agent maritime autonomy for unmanned surface vehicles in the swarm II harbor patrol demonstration," in *Unmanned Systems Technology XIX*, vol. 10195, SPIE, pp. 218–228, 2017.

[18] W. Wang, H. Zhang, Y. Li, Z. Zhang, X. Luo, and S. Xie, "USVS-SIM: A general simulation platform for unmanned surface vessels autonomous learning," *Concurrency and Computation: Practice and Experience*, vol. 34, no. 3, p. e6567, 2022.

[19] Z. He, C. Liu, X. Chu, W. Wu, M. Zheng, and D. Zhang, "Dynamic domain-based collision avoidance system for autonomous ships: Real experiments in coastal waters," *Expert Systems with Applications*, p. 124805, 2024.

[20] J. Liu, F. Yang, S. Li, Y. Lv, and X. Hu, "Testing and evaluation for intelligent navigation of ships: Current status, possible solutions, and challenges," *Ocean Engineering*, vol. 295, p. 116969, 2024.

[21] J. Hu, X. Yang, M. Lou, Q. Wang, H. Ye, and W. Liu, "Virtual-reality-based design and implementation with a parallel physical simulation platform for unmanned surface vehicle," in *Proceedings of International Conference on Cyber-Physical Social Intelligence*, Xi'an, China, pp. 262–267, 2023.

[22] Y. Dai, Y. He, X. Zhao, and K. Xu, "Testing method of autonomous navigation systems for ships based on virtual-reality integration scenarios," *Ocean Engineering*, vol. 309, p. 118597, 2024.

[23] P. Gomes, C. Silvestre, A. Pascoal, and R. Cunha, "A path-following controller for the DELFIMx autonomous surface craft," in *Proceedings of 7th IFAC Conference on Manoeuvring and Control of Marine Craft*, Lisbon, Portugal, pp. 1–6, 2006.

[24] C. Gentemann, J. P. Scott, P. L. Mazzini, C. Pianca, S. Akella, P. J. Minnett, P. Cornillon, B. Fox-Kemper, I. Cetinić, and T. M. Chin, "Saildrone: Adaptively sampling the marine environment," *Bulletin of the American Meteorological Society*, vol. 101, no. 6, pp. E744–E762, 2020.

[25] K. J. Casola, "System architecture and operational analysis of medium displacement unmanned surface vehicle Sea Hunter as a surface warfare component of distributed lethality," Ph.D. dissertation, Naval Postgraduate School, Monterey, USA, 2017.

[26] Miskovic, N., Nad, D. and Rendulic, I., "Tracking divers: An autonomous marine surface vehicle to increase diver safety," *IEEE Robotics & Automation Magazine*, vol. 22, no. 3, pp. 72–84, 2015.

[27] Grose, T.K., "Skip the skipper," *ASEE Prism*, vol. 31, no. 5, p. 8, 2022.

[28] Lambert, R., Page, B., Chavez, J. and Mahmoudian, N., "A low-cost autonomous surface vehicle for multi-vehicle operations," in *Proceedings of the Global Oceans 2020: Singapore–US Gulf Coast*, Biloxi, MS, USA, pp. 1–5, 2020.

[29] Sanchez, P.J.B., Marquez, F.P.G., Papaelias, M., Marini, S. and Govindaraj, S., "Enduruns project: Advancements for a sustainable offshore survey system using autonomous marine vehicles," in *Proceedings of the International Conference on Management Science and Engineering Management*, Ankara, Turkey, pp. 363–378, 2022.

[30] Sulligoi, G., Vicenzutti, A. and Menis, R., "All-electric ship design: From electrical propulsion to integrated electrical and electronic power systems," *IEEE Transactions on Transportation Electrification*, vol. 2, no. 4, pp. 507–521, 2016.

[31] Guo, J., Chen, Z. and Dai, Y., "Numerical study on self-propulsion of a waterjet propelled trimaran," *Ocean Engineering*, vol. 195, p. 106655, 2020.

[32] Wang, Z., Min, S., Peng, F. and Shen, X., "Comparison of self-propulsion performance between vessels with single-screw propulsion and hybrid contra-rotating podded propulsion," *Ocean Engineering*, vol. 232, p. 109095, 2021.

[33] Liu, J., Yan, X., Liu, C., Fan, A. and Ma, F., "Developments and applications of green and intelligent inland vessels in China," *Journal of Marine Science and Engineering*, vol. 11, no. 2, p. 318, 2023.

[34] Huntsberger, T., Aghazarian, H., Howard, A. and Trotz, D.C., "Stereo vision–based navigation for autonomous surface vessels," *Journal of Field Robotics*, vol. 28, no. 1, pp. 3–18, 2011.

[35] Vasilopoulos, V., Pavlakos, G., Schmeckpeper, K., Daniilidis, K. and Koditschek, D.E., "Reactive navigation in partially familiar planar environments using semantic perceptual feedback," *The International Journal of Robotics Research*, vol. 41, no. 1, pp. 85–126, 2022.

[36] Maersk, "Maersk selects Sea Machines for world's first AI-powered situational awareness system aboard a container ship," `https://www.marinelink.com/news/situational-technology436795`, 2018, [Online; accessed '7-July-2024].

[37] Rolls-Royce, "Intelligent awareness: Enhancing safety through intelligent data fusion," `https://www.ferryshippingsummit.com/wp-content/uploads/2018/03/Rolls-Royce.pdf`, 2018, [Online; accessed '7-July-2024].

[38] Kongsberg, "Intelligent awareness," `https://www.kongsberg.com/maritime/the-full-picture-magazine/2018/6/intelligent-awareness/`, 2019, [Online; accessed '7-July-2024].

[39] Zheng, H., Negenborn, R.R. and Lodewijks, G., "Survey of approaches for improving the intelligence of marine surface vehicles," in *Proceedings of the 16th International Conference on Intelligent Transportation Systems*, Den Haag, The Netherlands, pp. 1217–1213, 2013.

[40] Torben, T.R., Smogeli, Ø., Glomsrud, J.A., Utne, I.B. and Sørensen, A.J., "Towards contract-based verification for autonomous vessels," *Ocean Engineering*, vol. 270, p. 113685, 2023.

[41] Zheng, H., Li, J., Tian, Z., Liu, C. and Wu, W., "Hybrid physics-learning model based predictive control for trajectory tracking of unmanned surface vehicles," *IEEE Transactions on Intelligent Transportation Systems*, 2024.

[42] Newman, P.M., "Moos-mission orientated operating suite," Massachusetts Institute of Technology, Tech. Rep., vol. 2299, no. 08, 2008.

[43] Perez, T., Smogeli, O., Fossen, T.I. and Sorensen, A., "An overview of the marine systems simulator (MSS): a Simulink toolbox for marine control systems," *Modeling, Identification and Control*, vol. 27, no. 4, pp. 259–275, 2006.

[44] Skjetne, R., Fossen, T.I. and Kokotović, P.V., "Adaptive maneuvering with experiments for a model ship in a marine control laboratory," *Automatica*, vol. 41, no. 2, pp. 289–298, 2005.

[45] Transas, "WECDis equipment and marine simulators for naval training," `https://www.naval-technology.com/contractors/software/transas/`, 2024, [Online; accessed '7-July-2024].

[46] Kongsberg, "Autonomy is here – powered by KONGSBERG," `https://www.kongsberg.com/maritime/about-us/news-and-media/our-stories/autonomy-is-here--powered-by-kongsberg/`, 2024, [Online; accessed '9-July-2024].

[47] Proctor, A.A., Zarayskaya, Y., Bazhenova, E., Sumiyoshi, M., Wigley, R., Roperez, J., Zwolak, K., Sattiabaruth, S., Sade, H., Tinmouth, N., Simpson, B. and Kristoffersen, S.M., "Unlocking the power of combined autonomous operations with underwater and surface vehicles: Success with a deep-water survey AUV and ASV mothership," in *Proceedings of OCEANS*, Kobe, USA, pp. 1–8, 2018.

[48] Wolf, M.T., Rahmani, A., de la Croix, J.-P., Woodward, G., Vander Hook, J., Brown, D., Schaffer, S., Lim, C., Bailey, P., Tepsuporn, S. *et al.*, "CARACaS multi-agent maritime autonomy for unmanned surface vehicles in the swarm II harbor patrol demonstration," in *Unmanned Systems Technology XIX*, vol. 10195, SPIE, pp. 218–228, 2017.

[49] He, Z., Liu, C., Chu, X., Wu, W., Zheng, M. and Zhang, D., "Dynamic domain-based collision avoidance system for autonomous ships: Real experiments in coastal waters," *Expert Systems with Applications*, p. 124805, 2024.

[50] Colling, A., van Hassel, E., Hekkenberg, R., Flikkema, M., Nzengu, W., Ramne, B., van der Linden, E. and Kreukniet, N., "NOVIMAR: Deliverable 1.7: Vessel train handbook," University of Antwerp, Antwerp, Belgium, 2021.

[51] Hu, J., Yang, X., Lou, M., Wang, Q., Ye, H. and Liu, W., "Virtual-reality-based design and implementation with a parallel physical simulation platform for unmanned surface vehicles," in *Proceedings of International Conference on Cyber-Physical Social Intelligence*, Xi'an, China, pp. 262–267, 2023.

[52] Wu, W., Liu, C., Chu, X., Zhang, D., He, Z. and Zheng, M., "A novel ship short-term speed prediction method under the influence of currents," *Ocean Engineering*, p. 117847, 2024.

CHAPTER 7

Field Tests of USVs: Safe Navigation in the Lock Waterway

7.1 INTRODUCTION

$\mathbf{A}$ SHIP lock is a facility used for raising and lowering ships and other water-craft between low-level and high-level waterway, and connecting different water systems and improving waterway dimension and water flow condition, which has been the key node of inland river navigation [2]. In this chapter, we first analyze the requirements of safe navigation control in ship lock waterway, and then propose a systematic framework of multiple ship cooperative navigation, and finally demonstrate the effectiveness of the proposed navigation control methods and system in real scenarios.

7.2 NAVIGATION IN SHIP LOCK WATERWAY

The ship lock consists of navigation wall, lock chamber, gates, approach channels, cill, balance beam, paddle, winding gear, etc. [3]. In consideration of the lock's and ship's safety, a ship is required to keep low speed during passing through the lock. It is known that ship navigation in the lock waterway is restricted, and navigation risk, e.g., collision, becomes larger when the crew's driving level is declined. The low navigation efficiency in the lock waterway can lead to the ship congestion with the waterway transportation increasing. Therefore, the process for a ship to pass through the lock is one of the bottlenecks to improve the navigation ability of inland waterway. [4] pointed out that the traffic congestion at the Three Gorges Dam had become a serious problem in recent years. Two major problems are summarized for ship passing through the lock as follows.

- The efficiency is relatively low since two or more ships cannot move simultaneously with a close distance;

- The bollard is often damaged because ship cannot stop at the precise stop line because of human operational mistakes in the lock chamber.

There are several measures that can be taken to solve the above problems [1]. The first one is to deploy the standardization type ships to improve the ship total tonnage of single lock, but this will restrain the navigation of existing non-standard ships. The second one is to improve the traffic efficiency of the ship lock waterway with the traffic control methods based on the historical traffic data, e.g., see [5] and [4]. The last one is to form a ship formation to make the formation move integrally. A ship formation is fastened traditionally by the physical force, e.g., mooring rope, steel frame, but the physical way could consume much time, and the formation cannot change flexibly. A self-organizing formation can structure different types of ships without fastening physically, which means that the formation can be built quickly and change flexibly.

Cooperative Ship Formation Systems (CSFSs) or Cooperative Multi-Vessel Systems (CMVSs) have gained extensive attention recently [6, 7]. The CSFS can be widely used in practical scenarios, e.g., marine search, rescue mission, patrol, marine reconnaissance, commercial transportation [6]. Many studies for the CSFS focuses on proposing a new formation type or an improved control method with some general scenarios [8, 9, 10, 11, 12]. Building CSFSs is a useful way to solve the aforementioned problems in the ship lock waterway. There exist two problems to realize the cooperative control of ship formation in the lock waterway: one is that the lock navigation waterway is confined, which leads to the maneuverability in the lock waterway is different from that in open waterway, and the ship speed should be kept at the low level; the other is that many formation control methods of ship formation lack of the real experimental tests to verify the effectiveness of CSFSs and cooperative control methods. For most studies of formation control, in consideration of experimental complexity and safety, the effectiveness works of the proposed methods have been conducted by simulation, see [13, 14, 12, 15, 7, 16, 8]. [17] has developed a simulation platform for assessing cooperative adaptive cruise control. In comparison with other formation platforms, e.g. car and plane, the CSFS platform has more difficulties that mainly exist in the under-actuated defect for an individual ship and the unpredictability of the waterway. A real experimental platform of CSFC is necessary to validate the cooperative formation control methods and relative technologies. Perception of CSFS is the basis of meeting the requirements of precious speed and distance keeping control [18]. In the past, a ship did not have a requirement to transmit its navigation states to other ships and the shore center. However, for the CSFS, the ship navigation states, i.e., speed, position, rotation speed of main engines, relative distance and speed for each other, heading, should be shared precisely in real-time with the ships and shore center. An emergency switch way needs to be designed to guarantee the experimental safety in case of the failure of the CSFS. A model needs to be proposed to assess the performance of the CSFS control including the safety, reliability and efficiency.

From the literature, five structures of ship formation are mainly taken into account, i.e., leader-follower structure, behavioral-based structure, virtual structure, graph theory structure and distributed model predictive control (DMPC) structure.

For the leader-follower structure, the formation is required to be a designed formation shape and tracks with a leader [13, 19, 20]. For the behavioral-based structure, behavioral-based systems integrate several goal-oriented behaviors simultaneously, which can conduct multiple tasks such as obstacle avoidance, formation keeping at the same time [21, 22, 8]. For the virtual structure, this structure regards the entire formation as a single virtual rigid structure, which means individuals in the formation are similar to particles embedded in a rigid structure [23, 24, 25]. For the graph theory structure, a direct or un-directed graph is usually employed to model the communication topology [26, 12]. For the DMPC structure, the formation control problem is solved locally at the individual level with strong robustness even though the individual in the formation fails [15, 7, 16, 27]. Apart from the above classification method, according to the types of sensed and controlled variables, [28] categorized the formation control as position-based control, displacement-based control and distance-based control. The control of passing through the lock is more complicated than the formation control in the open waterway, to design an appropriate formation structure in the lock waterway is important to improve the navigation efficiency and safety of ships.

Based on the above requirements, the main objectives of this article concentrates on the CSFS structure and design, the study of cooperative formation control methods, implementation and experimental tests of CSFS. The rest of the article is organized as follows. Section 7.3 introduces the design and perception measures of CSFS. Section 7.4 proposes speed control, stop control and distance keeping control method for the leading ship and the following ships respectively. In Section 7.5, several tests are done with the real CSFS platform in Gezhouba Lock No.2 and No.3, and the results are analyzed to assess the control performance. Finally, Section 7.6 concludes main works and outlooks future plans.

7.3   SHIP NAVIGATION PERCEPTION IN COMPLEX WATERWAYS

The perception of ship formation in the lock waterway has its restrictions, which include lower GNSS signal strength, navigation wall and lock door interference, close distance between ships, etc. Hence, the specific structure and perception measures for the ship formation for passing through the lock are introduced in this section.

7.3.1   Passing through a lock

The structure of ship formation refers to the form of the construction and the operation of ship formation. It provides individual behavior and preset trajectory reference for the formation controller. As mentioned above, the typical ship formation structure includes leader-follower structure, virtual structure, behavior-based structure graph-based structure, etc. The study on the formation control structure is relatively mature, and each formation structure has an adaptive scenario. The ship formation passing through the lock is relatively stable and the ship can obtain the real-time feedback from the other ships. Meanwhile, the process of pass through a lock for ships is relatively complicated, which includes low-speed navigation in the navigation wall, entering the lock gate, navigation in the lock chamber, stop at the lock chamber, leaving the lock chamber, etc. To improve the formation safety and efficiency, a leader-follower structure combined with the behavior-based structure is designed in this article.

To ensure the safety of navigation, the speed of ships passing through the lock is low, and the ships should meet the navigation rule requirements. An improved leader-follower structure is proposed to organize the ship formation through the lock, that is, the ship formation is composed of one leading ship and several following ships, as shown in Fig. 7.1. When a ship stops at the stop line in the lock chamber, the headline of the ship cannot be allowed to cross the forbidden line. The leading ship sets its speed according to the navigation rules of the lock waterway. Each following ship has its own target ship that could be the leading ship or another following ship, which means a following ship can have two roles, i.e., the following ship and the target ship. The following ships dynamically adjust their speed according to the relative distance and relative speed to the leading ship. The leading ship and the following ships have relevant different behaviors. For instance, the leading ship has the speed control behavior, stop control behavior, stop-finished behavior and emergency stop behavior, and the following ship has distance keeping behavior, stop-finished behavior and emergency stop behavior. Hence, the proposed formation structure is the combination of leader-follower and behavior-based structure. The detail of the formation control will be discussed in Section 7.4.

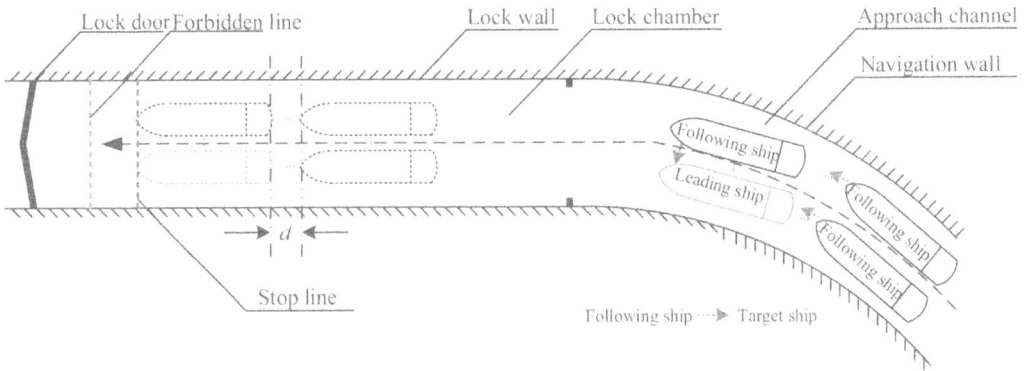

Figure 7.1: Formation structure of CSFS.

7.3.2 Perception measurements in the lock waterway

The ship formation perception system is an important part of CSFS. The information by the perception system contains ship navigation environment status, own ship's and other ships' navigation status, lock status, monitoring orders, etc. The perception scheme illustrates the perception information flow and the perception measures, which is shown in Fig. 7.2.

1. Environmental perception based on millimeter-wave radar

    In consideration of relatively short ranging ability (less than 300 m), a MilliMeter-Wave (MMW) radar is scarcely utilized for ship environment

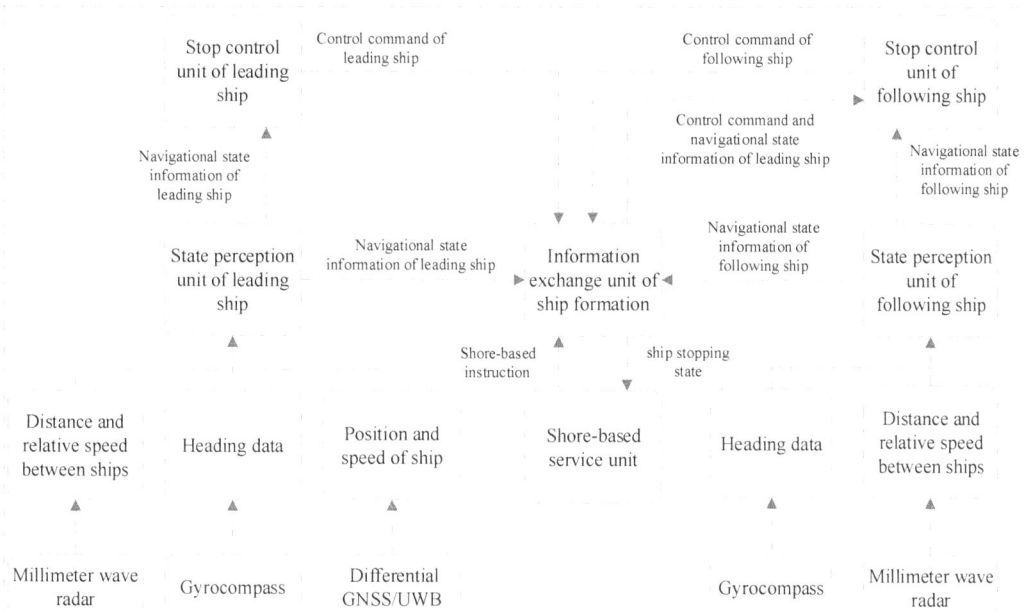

Figure 7.2: Perception scheme of CSFS.

perception. Considering that the ship formation in the lock waterway is tight, the relative distance between a ship and its around ship is usually less than 100 m, an MMW-based perception measure is adopted in this article. Explicitly, the environmental perception system consists of an MMW radar (ARS408-21 is used in this article) mounted at the ship bow, and three short-distance MMW radar (CAR28F is used in this article) mounted at the ship stern, port side and starboard side. An MMW radar operates in the MMW frequency band (30–300 GHz). The MMW radar has the advantages of both microwave and photoelectricity, and has small size, light weight, and high spatial resolution [29]. Compared with the solid-state radar and laser radar, the MMW radar has a more substantial penetration of fog, smoke, dust, and others. Moreover, the MMW radar has a unique advantage in the dynamic perception of ship navigation. The MMW radar can obtain the precise distance between ships and the relative speed of ships. In order to prevent the MMW radar from sweeping to both sides of the lock wall when the ship is in the lock chamber, which will affect the ship distance data, the effective sweep angle of the MMW radar is set to filter the data to eliminate the influence of the lock wall on the target measurement. Taking the front MMW radar of the ship passing through the lock as an example, the radar scanning angle setting is shown in Fig. 7.3.

As is shown in Fig. 7.3, φ is the effective sweep angle of MMW radar, b is the width of the lock chamber; c is a shorter distance between the ship and the lock wall; d is the shortest allowed distance between adjacent ships in the longitudinal direction; θ is the angle between the ship heading and the lock chamber direction. To avoid the lock wall affects the distance measurement of

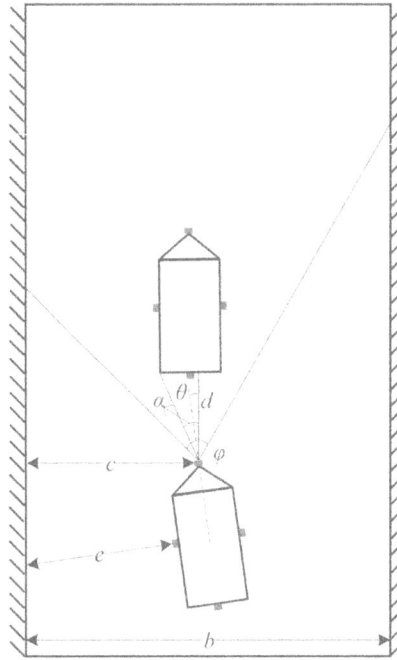

Figure 7.3: Scan angle setting of MMW radar.

d, the sweep angle of the MMW radar needs to be set as follows in consideration of the ship heading deflection:

$$d \le \frac{c}{\sin\left(\frac{\varphi}{2} + \theta\right)}, \tag{7.1}$$

where the value of c can be measured by the port or starboard MMW radar. According to the MMW radar ranging value, the precise position coordinates of the ship in the lock chamber can be calculated. As is shown in Fig. 7.4, $ABDC$ represents the front ship, $EFHG$ represents the object ship, and $IJLK$ represents the rear ship. Assuming that the front and rear ship are both a rectangular target with a fixed size in general. The distances from $ABDC$ to the left and right lock wall are a_1 and a_2, respectively. The distances from $EFHG$ to $ABDC$, the left wall, the right wall, and $IJLK$ are d_1, d_2, d_3 and d_4, respectively. The distances from $IJLK$ to the left and right lock wall are f_1 and f_2, respectively. Define the included angle of $EFHG$ to the forward direction of the lock chamber is ψ. The positioning parameters are calculated as follows:

$$a_1 = b_1 + \frac{W}{2}(\cos\psi - 1) + d_1\sin\psi \tag{7.2}$$

$$\begin{cases} b_1 = d_2\cos\psi + (L/2 - x_2)\sin\psi \\ b_2 = d_3\cos\psi - (L/2 - x_3)\sin\psi \\ b_3 = b_1 - L\sin\psi \\ b_4 = b_2 + L\sin\psi \end{cases} \tag{7.3}$$

$$\begin{cases} c_1 = d_1 \cos \psi - \frac{W}{2} \sin \psi \\ c_2 = L \cos \psi \\ c_3 = \frac{W}{2} \sin \psi + d_4 \cos \psi \end{cases} \tag{7.4}$$

where W and L are the ship width and length of $EFHG$.

To eliminate the interferences of the environmental objects, such as water surface, lock wall and so on, several filtering measures are utilized. Specifically, most of the lock wall and lock door interferences can be eliminated with the appropriate sweep angle setting. On the basis of the differences between the ship object and the water surface, the interference elimination of the water surface can be realized by setting the value range of the reflectivity and the area of Radar Cross Section (RCS).

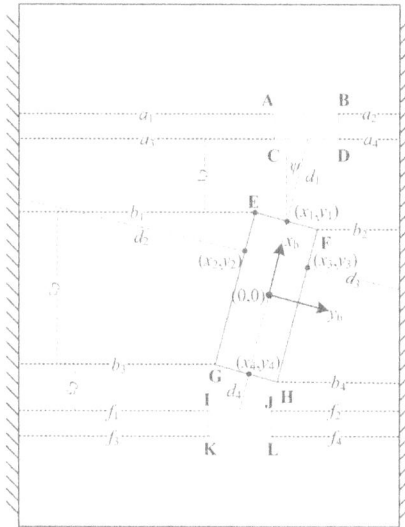

Figure 7.4: Ship position calculation.

2. Ship status perception

   The gyrocompass is used to obtain the ship's heading and attitude data. Differential GNSS is utilized to obtain the accurate real-time position, speed, and heading data of the ship. Ultra Wide Band (UWB) positioning system is a local positioning system, which can keep the positioning ability even if GNSS signal is weak or lost. The data acquisition terminal is used to obtain the data of ship main engine and propeller speeds, and integrate all the ship status data. The onboard computer is used for the data processing, data analysis, algorithm realization, and status displaying and saving. The ship state perception hardware architecture is shown in Fig. 7.5.

7.3.3   Ship and shore interaction

CSFS consists of one shore-based system and several ship onboard systems. Ship and shore interaction mainly realizes ship-to-shore communication, shore-based monitoring, data storage, human-machine interaction. Ship-to-shore interaction data mainly

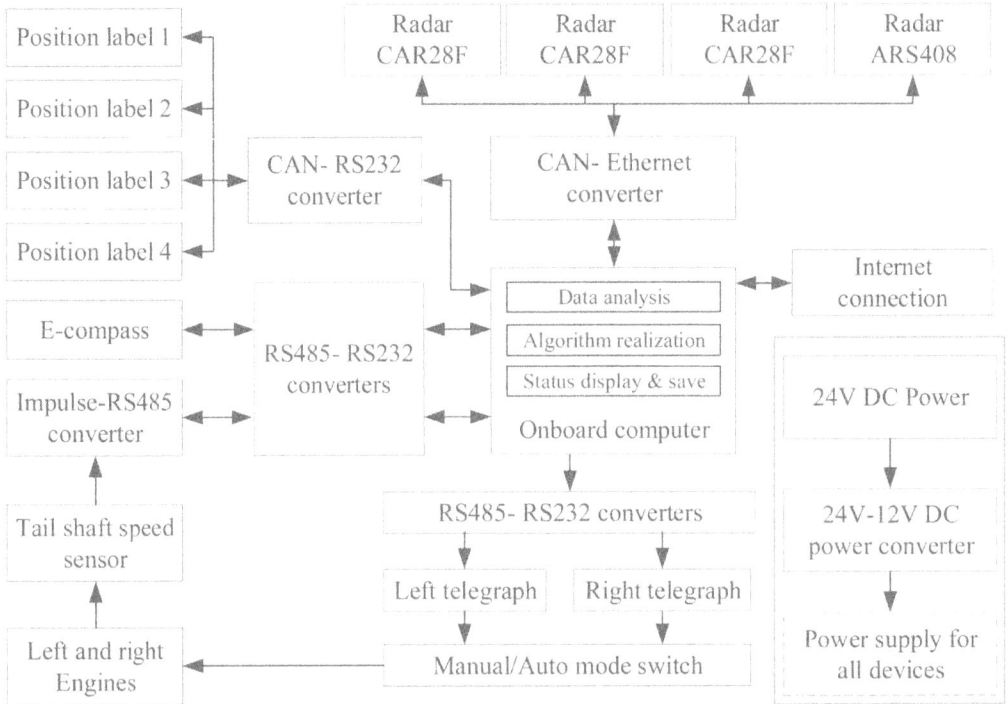

Figure 7.5: Hardware architecture of ship status system.

include ship pose positioning data, MMW radar data, lock status data, lock signal status data, monitoring information data, etc. The information interaction is illustrated in Fig. 7.6.

7.4 CONTROL METHODS FOR PASSING THROUGH THE LOCK COOPERATIVELY

Navigation behaviors of a leading ship for passing through a lock correspond to four control mode, i.e., speed control mode, stop control mode, stop-finished mode and emergency stop control mode. The control process of the leading ship is shown in Fig. 7.7. Similarly, navigation behaviors of a following ship for passing through a lock correspond to three control modes, i.e., distance keeping control mode, stop-finished mode and emergency stop control mode. The control process of the following ship is shown in Fig. 7.8. The mode switch procedure and longitude distance conversion for the leading ship and the following ship are illustrated as Fig. 7.9.

From Fig. 7.9, it can be seen that the actual trajectory in the lock waterway is transferred to the longitude distance from the starting point. In the coordination of longitude distance and speed axis, the navigation behavior is described as the behavior state (s_x, v_x), where s_x is the longitude distance, v_x is the ship speed. For instance, the state $(s_P, 0)$ stands for the stop-finished behavior or mode.

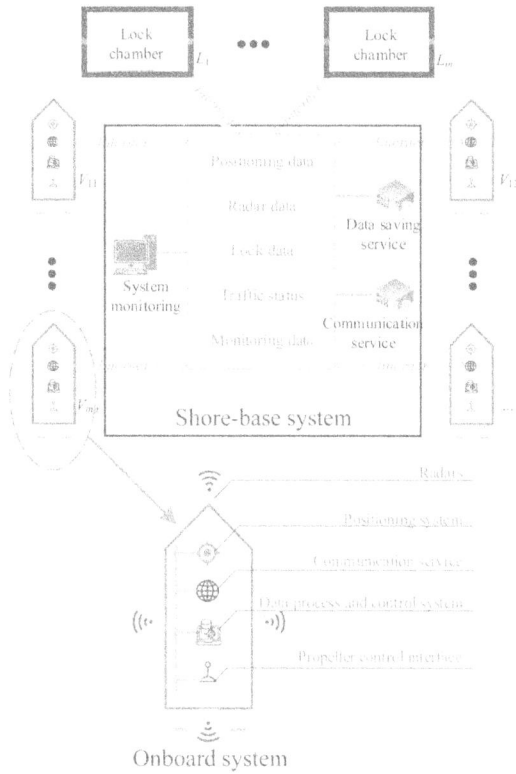

Figure 7.6: Information interaction between ships and shore

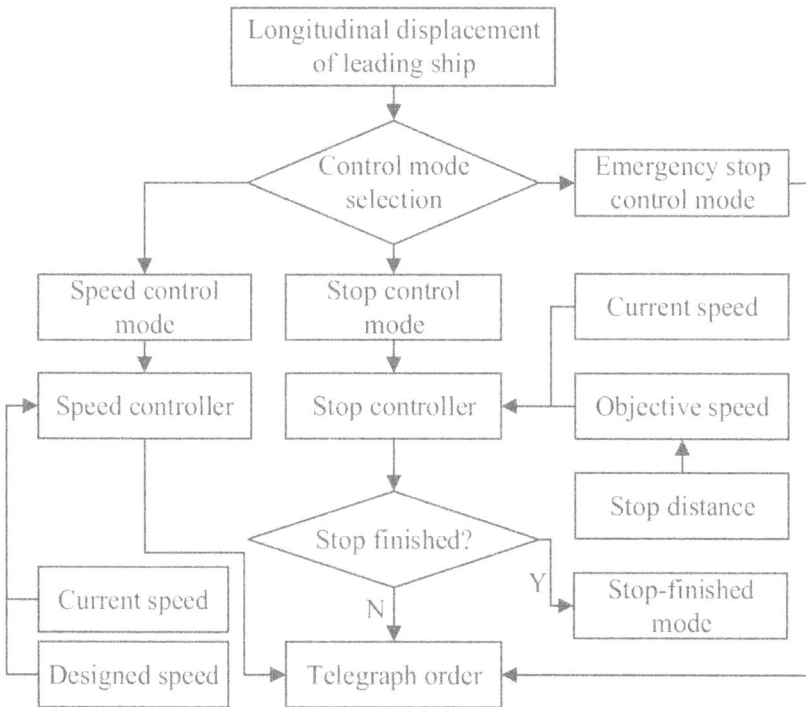

Figure 7.7: Leading ship control principle.

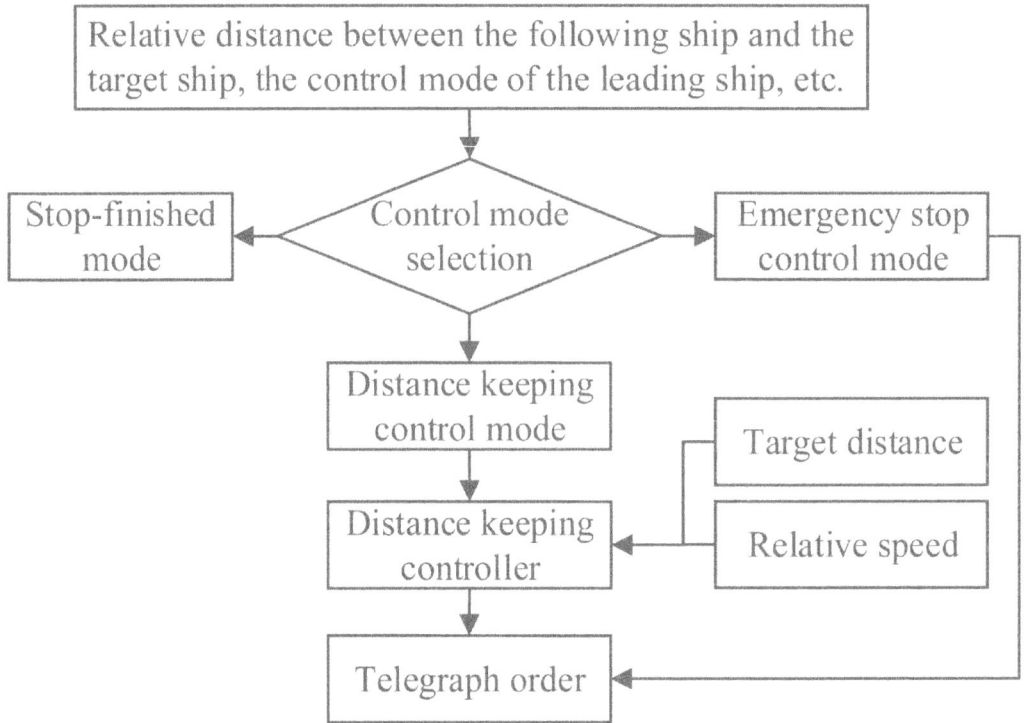

Figure 7.8: Following ship control principle.

7.4.1 Speed control method

Speed control refers to the automatic control of the longitudinal speed of the leading ship in accordance with the navigation rules of the lock water area. This article uses the fuzzy Proportion Integral Differential (PID) method to realize the speed control of the ship. The target speed is defined as v_0. Take the telegraph order as the system input $u(t)$ at time t.

$$u(t) = K_\mathrm{P}e(t) + K_\mathrm{I}\sum_{i=0}^{t} e(i) + K_\mathrm{D}[e(t) - e(t-1)], \tag{7.5}$$

where K_P, K_I and K_D are the PID parameters. $e(t)$ is the control error and represented as $e = v(t) - v_0$ at time t, in which $v(t)$ is the measured speed of the ship and v_0 is the objective speed of the ship. Considering that the forward and reverse efficiency of ship propellers at the same rotation speed are different, PID parameters should be set respectively. To prevent manipulating the ship telegraph too frequently, a fuzzy speed zone is proposed, namely:

$$u(t) = \begin{cases} 0, & \text{if } v(t) \in [-v_\mathrm{f}, v_\mathrm{f}] \\ K_\mathrm{P}e(t) + K_\mathrm{I}\sum_{i=0}^{t} e(i) + K_\mathrm{D}[e(t) - e(t-1)], & \text{else} \end{cases}, \tag{7.6}$$

where $v_\mathrm{f} > 0$ is the preset fuzzy speed.

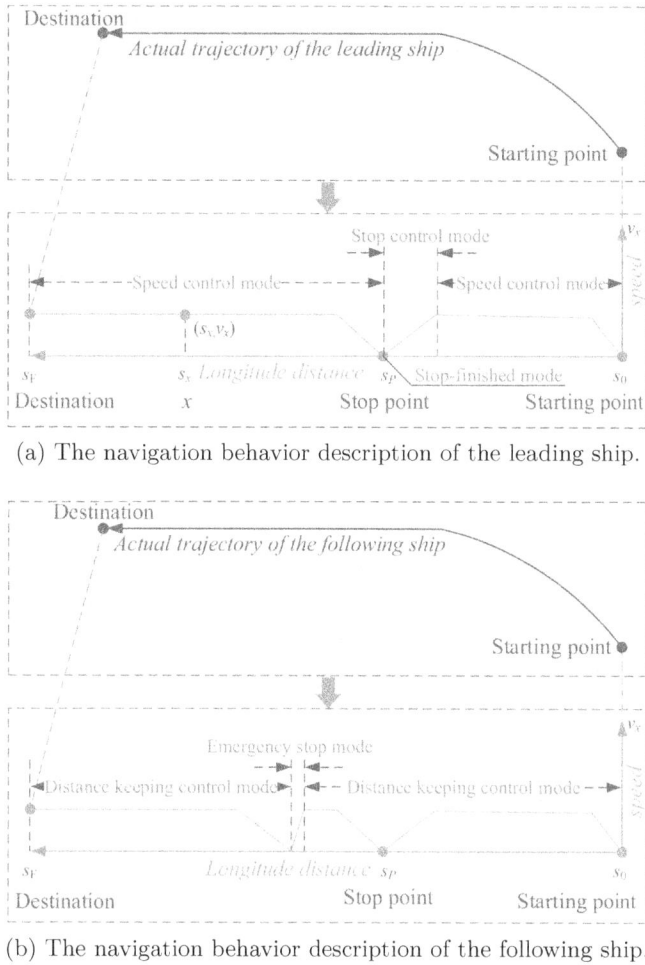

(a) The navigation behavior description of the leading ship.

(b) The navigation behavior description of the following ship.

Figure 7.9: Behavior description of the formation ship during passing through the lock.

7.4.2 Stopping control method

Accidents such as damage to the bollards of the lock chamber and collision of the lock gate or wall caused by the ship's mishandling often occur. Hence, it is essential to realize automatic and precise stop control in the lock chamber. When a leading ship in the formation passing through the lock stops at the designated stop position in the lock chamber, the following ship will keep a safety distance from the leading ship and stop accordingly. The navigation longitudinal distance for the whole stop procedure is defined as the stop distance, denoted as d_s. Stop distance is related to the ship maneuverability, the ship speed and telegraph input. To make the ship stop accurately in the lock chamber, an appropriate target stop distance should be set as d_0. A ship stopping at the stop line means that the ship's speed keeps 0 and the ship's telegraph keeps neutral (no power output for the propellers). Actually, the process of stopping a ship can be described as the process of controlling the ship's motion state $(v(t), u(t))$ to $(0,0)$. Among the states, $u > 0$ is ahead telegraph order, $u < 0$ is astern telegraph order, $u = 0$ is neutral telegraph order. The target speed function f_s during the ship stopping in the lock chamber is defined as:

$$v_0(t) = f_s(d_t) = \frac{d_t}{d_{s0}}, \tag{7.7}$$

where d_t is the distance between the ship and the stop line at time t, and when the ship crosses the stop line, we have $d_t < 0$. $v_0(t)$ is the ship's target speed. d_{s0} is the distance to judge the speed control mode or the stop control mode for a leading ship. The stop distance interval is set to judge whether the stop procedure is finished or not, namely:

$$\textit{Leading ship control mode} = \begin{cases} \textit{Speed control mode}, & \text{if } d_t \in (d_{s0}, \infty) \\ \textit{Stop control mode}, & \text{if } d_t \in (d_{se}, d_{s0}) \\ \textit{Stop-finished mode}, & \text{if } d_t \in [-d_{se}, d_{se}) \\ \textit{Emergency stop mode}, & \text{if } d_t \in (-\infty, -d_{se}) \end{cases}. \tag{7.8}$$

where d_{se} is the fuzzy distance of the stop-finished mode.

7.4.3 Distance keeping control method

Each following ship i has its own target ship that could be another following ship or the leading ship. The following ship keeps the relative distance and speed from the target ship according to the target ship's speed and distance. The distance between the following ship i and the target ship is set as the control feedback, i.e., the system output, which is denoted as $d_i(t)$ at time t. The telegraph is set as the control output, i.e., the system input, which is denoted as $u_i(t)$ at the time t. The target keeping distance is set as the control objective, which is denoted as d_0. A fuzzy PID method is used as the distance keeping control method for the following ship i.

$$u_i(t) = K_{iP}e_i(t) + K_{iI}\sum_{i=0}^{t} e_i(i) + K_{iD}[e_i(t) - e_i(t-1)], \tag{7.9}$$

Considering that the ship propeller forward and reverse efficiency at the same speed are different, PID parameters, i.e., K_{iP}, K_{iI}, and K_{iD}, should be set for the ahead and astern telegraph order, respectively. Assuming that the relative velocity of the two ships is $v_{ir}(t)$. $v_{ir}(t) < 0$ indicates the relative distance of two ships is shrinking. In order to maintain the stability of ship tracking, the tracking target distance fuzzy area is set as $[d_{i0} - d_{if}, d_{i0} + d_{if}]$. The fuzzy area $u(t)$ is considered as:

$$u(t) = \begin{cases} K_{ir}v_{ir}(t), & \text{if } d_i(t) \in [d_{i0} - d_{if}, d_{i0} + d_{if}] \\ K_{iP}e(t) + K_{iI}\sum_{i=0}^{t} e_i(i) + K_{iD}[e_i(t) - e_i(t-1)], & \text{else} \end{cases}, \tag{7.10}$$

where, $d_{if} \geq 0$ is the fuzzy distance threshold value and K_{ir} represents the control parameter. The distance $d_i(t)$ and relative speed $v_{ir}(t)$ between the following ship i and the target ship are measured by MMW radars.

7.5   CSFS EXPERIMENTAL TESTS

To test the effectiveness of the designed CSFS and the proposed formation control methods in the real lock waterway, several tests in lock No. 2 and No. 3 of Gezhouba Lock, China have been conducted. The CSFS is built, of which the main devices are shown Fig. 7.10.

Figure 7.10: Main devices of built CSFS.

The test sites and scenarios are shown in Fig. 7.11. Chamber of Lock No. 2 is 280 m long and 34 m wide. Chamber of Lock No. 3 is 120 m long and 18 m wide. Several experimental tests, including distance keeping control, speed control, and lock crossing control of the two-ship formation were completed. See Table 7.1 for details. The main parameters of the scenario and controller are set as Table 7.2.

TABLE 7.1   Experimental test arrangement.

Date	Site	Item	Ships
Jun. 26, 2019	No. 3 lock	distance keeping control	*Haixun 12911* and *Haixun 12907*
Sep. 5, 2019	No. 2 lock	speed control	*Hanggong 201*
Nov. 9, 2019	No. 2 lock	cooperative control	*Hangong 201* and *Hanggong 07*

7.5.1   Distance keeping control test

During the distance keeping control test, *Haixun 12907* acts as a leading ship (manual control) and *Haixun 12911* acts as a following ship (automatic control). To verify the performance of *Haixun 12911*'s distance keeping in the scenario of the change of relative speed between *Haixun 12907* and *Haixun 12911*. The designed keeping

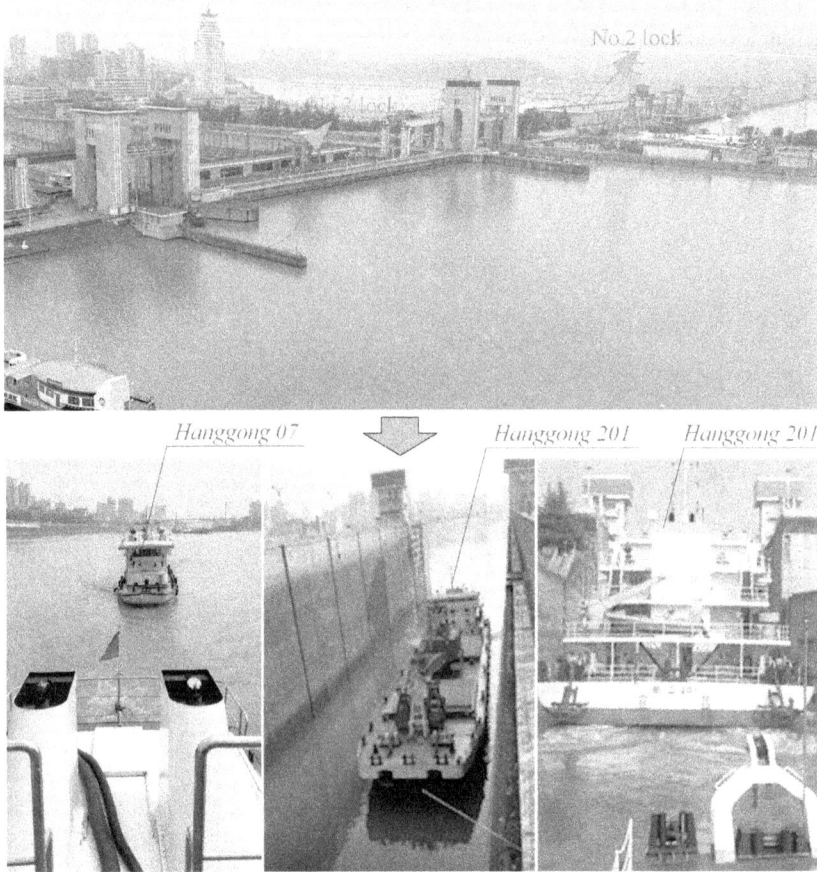

Figure 7.11: Test sites and scenarios in Gezhouba lock.

distance is set at 20.0 m, and the *Haixun 12907* (the leading ship) is manipulated at around 1.0 m/s. To denote the telegraph order clearly, the telegraph order range is set to (-100,100), where -100 and 100 stand for the full astern and the full ahead order, respectively. The test results are shown in Fig. 7.12. According to the experimental data, the average tracking distance error is 2.43 m, and the average relative speed is 0.33 m/s. From the Fig. 7.12c and 7.12d, it can be seen that the propeller rotation speed change is consistent with the telegraph input value. The distance keeping control results show that the proposed distance keeping control method can meet the requirement of the ship formation control in the lock waterway.

7.5.2 Speed control test

In the scenario of speed control test, a leading ship, i.e., *Hanggong 201*, is controlled by the CSFS. The ship initial navigation speed is set as 0.8 m/s. The test results are shown in Fig. 7.13. From the experimental data, the average speed control error

TABLE 7.2 Parameter setting for the test scenario and the controller.

Parameters	Value
Forward PID parameter of distance keeping control	$K_P = 0.015, K_I = 0, K_D = 0.002$
Reverse PID parameter of distance keeping control	$K_P = 0.060, K_I = 0, K_D = 0.002$
Forward PID parameter of speed control	$K_P = 0.30, K_I = 0, K_D = 0.02$
Reverse PID parameter of speed control	$K_P = 0.60, K_I = 0, K_D = 0.02$
Fuzzy distance threshold d_f	3.00 m
Fuzzy speed threshold v_f	0.40 m/s
Stop control parameter d_{se}	3.00 m
K_r of forward and reverse telegraph	0.20, 0.40
Stop distance setting d_{s0}	80.0m
Forward and reverse power proportion of full power for *Haixun 12911*	7%, 20%
Forward and reverse power proportion of full power for *Hanggong 201*	10%, 100%
Forward and reverse power proportion of full power for *Hanggong 07*	10%, 30%

(a) Tracking distance change during distance keep-(b) Relative speed between the leading ship and
ing   following ship

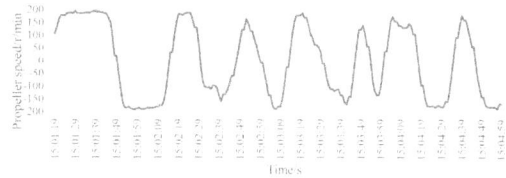

(c) Telegraph input (-100 – 100) of the following (d) Propeller rotation speed of the following ship
ship

Figure 7.12: Distance keeping control of the following ship.

is 0.08 m/s. From the speed control test results, it can be seen that the ship actual speed approaches to the designed speed within a short time and keeps at the value around the designed speed by the use of the proposed speed control method.

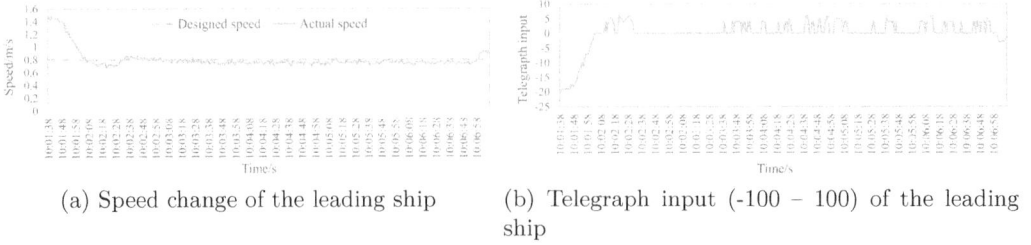

(a) Speed change of the leading ship

(b) Telegraph input (-100 – 100) of the leading ship

Figure 7.13: Speed control of the leading ship.

7.5.3 Cooperative control test

During the cooperative control test of the two-ship formation, the leading ship *Hanggong 201* is automatically controlled by CSFS during the speed control and stop control mode. The following ship *Hanggong 07* is also automatically controlled by CSFS during the distance keeping control mode and stop control mode. The data of cooperative control process is shown in Fig. 7.14. After entering the stop-ship mode, the speed slowly drops to 0. At this time, the distance between the leading ship and the set stop line is 2.0 m. After starting again, the leading ship's speed steadily accelerates to 0.8 m/s, as shown in Fig. 7.14a and 7.14b. The following ship *Hanggong 07* maintains at the speed below 1.5 m/s during the upstream crossing of the lock, and the longitudinal distance from the leading ship is kept at the set distance of about 30 m, as shown in Fig. 7.14c and 7.14d. Due to the longitudinal in collinearity between the leading ship and the following ship when mooring, the MMW radar measurement output is 0. The CSFS can realize ship speed control, ship distance keeping control, and ship stop control from the analysis of the test results.

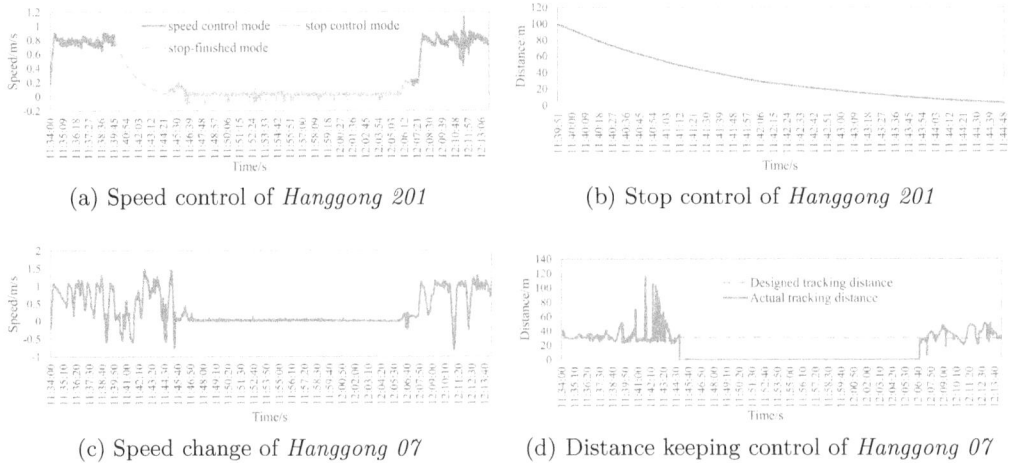

(a) Speed control of *Hanggong 201*

(b) Stop control of *Hanggong 201*

(c) Speed change of *Hanggong 07*

(d) Distance keeping control of *Hanggong 07*

Figure 7.14: Results of cooperative control during passing through the lock.

7.6 CONCLUSIONS

To improve the safety and efficiency when ships pass through the lock waterway, a CSFS is designed and implemented in Gezhouba Lock waterway. MMW radars are utilized to measure the relative distance and the relative speed between ships. Cooperative formation control methods are proposed to control ships passing through the lock automatically. The main conclusions are as follows:

1. The ship formation structure based on the leading-following and behavior is designed. The leading ship has speed control mode, stop control mode, stop finished mode and emergency stop mode. The following ship has distance keeping control, stop control mode and emergency stop mode.

2. A ship status perception method based on MMW radar and ship-shore interaction are proposed to achieve the acquisition of the ship's navigation states and formation states in the confined lock waterway environment.

3. The cooperative speed control methods of ship formation, i.e., distance keeping control, speed control and stop control, is proposed. CSFS and control methods have been tested several times in the waterway of Gezhouba Lock No. 2 and No. 3. The average tracking error of the distance keeping control for the following ship is 2.43 m, and the average relative speed for the leading ship is 0.33 m/s. Speed control error for the leading ship is 0.08 m/s. For the stop control, the leading ship and the following can complete the accurate stop control.

In future research, the distributed control idea can be considered to realize the cooperative control of multiple ships passing through the lock to improve the formation system's real-time performance and robustness. The cooperative formation control method can also be applied to multi-stage lock formation scheduling and control to solve the optimal control among multiple formations and improve the navigation efficiency of ships in the whole lock waterway.

Bibliography

[1] C. Liu, J. Qi, X. Chu, M. Zheng, W. He, "Cooperative ship formation system and control methods in the ship lock waterway," *Ocean Engineering*, vol. 226, p. 108826, 2021.

[2] V. Bugarski, T. Bačkalić, U. Kuzmanov, "Fuzzy decision support system for ship lock control," *Expert Systems with Applications*, vol. 40, no. 10, pp. 3953–3960, 2013.

[3] H. Wang, Z. Zou, "Numerical study on hydrodynamic interaction between a berthed ship and a ship passing through a lock," *Ocean Engineering*, vol. 88, pp. 409–425, 2014.

[4] X. Zhao, Q. Lin, H. Yu, "A co-scheduling problem of ship lift and ship lock at the Three Gorges Dam," *IEEE Access*, vol. 8, 2020.

[5] T. Backalic, M. Bukurov, "Modelling of the ship locking process in the zone of ship lock with two parallel chambers," *Annals of the Faculty of Engineering Hunedoara*, vol. 9, no. 1, pp. 187–192, 2011.

[6] W. Xie, B. Ma, T. Fernando, H.H.C. Iu, "A new formation control of multiple underactuated surface vessels," *International Journal of Control*, vol. 91, no. 5, pp. 1011–1022, 2018.

[7] L. Chen, H. Hopman, R.R. Negenborn, "Distributed model predictive control for vessel train formations of cooperative multi-vessel systems," *Transportation Research Part C: Emerging Technologies*, vol. 92, pp. 101–118, 2018.

[8] F. Arrichiello, S. Chiaverini, T.I. Fossen, "Formation control of underactuated surface vessels using the null-space-based behavioral control," In *Proceedings of the 2006 IEEE/RSJ International Conference on Intelligent Robots and Systems*, 2006, pp. 5942–5947.

[9] T. Li, R. Zhao, C.L.P. Chen, L. Fang, C. Liu, "Finite-time formation control of under-actuated ships using nonlinear sliding mode control," *IEEE Transactions on Cybernetics*, vol. 48, no. 11, pp. 3243–3253, 2018.

[10] K.D. Do, "Practical formation control of multiple underactuated ships with limited sensing ranges," *Robotics and Autonomous Systems*, vol. 59, no. 6, pp. 457–471, 2011.

[11] F. Fahimi, "Non-linear model predictive formation control for groups of autonomous surface vessels," *International Journal of Control*, vol. 80, no. 8, pp. 1248–1259, 2007.

[12] C. Huang, X. Zhang, G. Zhang, "Improved decentralized finite-time formation control of underactuated usvs via a novel disturbance observer," *Ocean Engineering*, vol. 174, pp. 117–124, 2019.

[13] Z. Sun, G. Zhang, Y. Lu, W. Zhang, "Leader-follower formation control of underactuated surface vehicles based on sliding mode control and parameter estimation," *ISA transactions*, vol. 72, pp. 15–24, 2018.

[14] B.S. Park, "Adaptive formation control of underactuated autonomous underwater vehicles," *Ocean Engineering*, vol. 96, pp, 1–7, 2015.

[15] H. Zheng, J. Wu, W. Wu, R.R. Negenborn, "Cooperative distributed predictive control for collision-free vehicle platoons," *IET Intelligent Transport Systems*, vol. 13, no. 5, pp. 816–824, 2018.

[16] S. Wei, Y. Chai, B. Ding, "Distributed model predictive control for multiagent systems with improved consistency," *Journal of Control Theory and Applications*, vol. 8, no. 1, pp. 117–122, 2010.

[17] L. Cui, J. Hu, B.B. Park, P. Bujanovic, "Development of a simulation platform for safety impact analysis considering vehicle dynamics, sensor errors, and communication latencies: assessing cooperative adaptive cruise control under cyber attack," *Transportation research part C: emerging technologies*, vol. 97, pp. 1–22, 2018.

[18] Y. Xie, X. Yang, F. Xun, L. Wang, "Integrated data acquisition terminal used on board," in *Proceedings of the 18th International Symposium on Communications and Information Technologies (ISCIT)*, Bangkok, Thailand, 2018, pp. 127–130.

[19] K. Shojaei, "Leader-follower formation control of underactuated autonomous marine surface vehicles with limited torque," *Ocean Engineering*, vol. 105, pp. 196–205, 2015.

[20] S. He, M. Wang, S. Dai, F. Luo, "Leader-follower formation control of usvs with prescribed performance and collision avoidance," *IEEE Transactions on Industrial Informatics*, vol. 15, no. 1, pp. 572–581 , 2018.

[21] T. Balch, R.C. Arkin, "Behavior-based formation control for multirobot teams," *IEEE Transactions on Robotics and Automation*, vol. 14, no. 6, pp. 926–939, 1998.

[22] J.R.T. Lawton, R.W. Beard, B.J. Young, "A decentralized approach to formation maneuvers," *IEEE transactions on robotics and automation*, vol. 19, no. 6, pp. 933–941, 2003.

[23] M.A. Lewis, K.H. Tan, "High precision formation control of mobile robots using virtual structures," *Autonomous robots*, vol. 4, no. 4, pp. 387–403, 1997.

[24] H. Mehrjerdi, J. Ghommam, M. Saad, "Nonlinear coordination control for a group of mobile robots using a virtual structure," *Mechatronics*, vol. 21, no. 7, pp. 1147–1155, 2011.

[25] Q. Qin, T. Li, C. Liu, C.L.P. Chen, M. Han, "Virtual structure formation control via sliding mode control and neural networks," in *Proceedings of the 2017 International Symposium on Neural Networks*, 2017, pp. 101–108..

[26] B.S. Park, S.J.!Yoo, "An error transformation approach for connectivity preserving and collision-avoiding formation tracking of networked uncertain underactuated surface vessels," *IEEE Transactions on Cybernetics*, vol. 49, no. 8, pp. 2955–2966, 2018.

[27] H. Zheng, R.R. Negenborn, G. Lodewijks, "Fast admm for distributed model predictive control of cooperative waterborne AGVs," *IEEE Transactions on Control Systems Technology*, vol. 25, no. 4, pp. 1406–1413, 2016.

[28] K.K. Oh, M.C. Park, H.S. Ahn, "A survey of multi-agent formation control," *Automatica*, vol. 53, pp. 424–440, 2015.

[29] K. Yoneda, N. Hashimoto, R. Yanase, M. Aldibaja, N. Suganuma, "Vehicle localization using 76GHz omnidirectional millimeter-wave radar for winter automated driving," in *Proceeding of the 2018 IEEE Intelligent Vehicles Symposium*, 2018, pp. 971–977.

CHAPTER 8

Field Tests of USVs: Ship Turning Maneuver Tests

8.1 INTRODUCTION

SHIP TRAJECTORY prediction is the basis of ship safe and precise navigation control [1]. In this chapter, we firstly build an extreme short-term prediction model based on the trajectory data collected during a ship turning test under the influence of sea currents. We derive the direction and velocity of sea currents based on the velocity characteristic distribution from actual turning test data. The ship trajectory extreme short-term prediction model is built by integrating the sea current information and ship response model, and the sea current information is simultaneously online corrected based on the prediction errors. Finally, the prediction accuracy is tested and verified in comparison with other algorithms with the actual ship turning test data.

8.2 MODELING OF SHIP EXTREME SHORT-TERM TRAJECTORY PREDICTION

Ship maneuverability performance is closely related to ship navigation safety (e.g., collision avoidance [2] and energy saving [3], which is usually reflected in turning ability, yaw-checking ability, course-keeping ability, stopping ability, etc. [4]. Ship motion tests are utilized to evaluate the maneuverability and can be roughly classified to real ship tests [5], model ship tests and numerical simulation tests [6, 7]. The real ship tests generally involve the Zigzag, turning, and ship-stopping tests in the open and calm waters. The model ship test is generally conducted under static and dynamic planar motion mechanisms using a scale model ship in the ship basin environment or a self-propelled scale model ship test in calm water. The numerical simulation tests use computational fluid dynamics methods to simulate the maneuverability of the scale model ship. To sum up, the real ship tests are mostly credible, especially with the change of ship hull surface, draught, or mechanism, but with higher requirements of test condition and cost. The model ship and numerical simulation tests have lower conditions, of which accuracy is hard to be guaranteed [8].

DOI: 10.1201/9781003660590-8

The ship turning test is used to evaluate the ship steering performance and can provide important security of the ship navigation safety throughout its life cycle [9]. Conventional ship turning tests are usually carried out in calm water, with relatively little consideration of environmental factors, such as wind, waves, and current [8]. However, as the maneuverability tests are carried out in coastal and oceanic waters, the ship turning trajectory would be irregular owing to the influence of environmental disturbances [10]. Therefore, it is necessary to analyze the influence of environments on the ship turning motion, build the ship motion model with the environmental disturbances, and achieve a satisfactory prediction of the ship motion trajectory. In terms of ship trajectory prediction, the influence of waves on ship longitudinal movement, lateral drift, and turning was studied in [11], and achieved a prediction of ship turning trajectory under the influence of waves through numerical simulation. In [12], a Norrbin model of ship motion considering the effects of wind, waves, and tidal current was built and achieved satisfactory turning trajectory prediction under simulation conditions. For the ship trajectory prediction under real sea conditions, in [13], a deep learning network has been used to identify environmental parameters, i.e., wind, waves, and current, based on the results of the JARI unmanned boat turning test conducted in the open sea. The predicted results based on the turning test agreed with the real test trajectory. However, the deep learning algorithms lack interpretability and adaptability to different environments, which needs further verification.

The ship turning test is generally used to identify a response model that takes rudder angle as an input and takes heading as an output [14]. The identified response model can be usually used for the ship short-term trajectory prediction with known velocity [9]. In [15], the Long Short-Term Memory (LSTM) neural network method and Gaussian Process Regression method have been used to achieve a short-term prediction of ship motion, attaining the over 50 s of prediction time for rolling, pitching, and heaving motion. When a ship sails in the open sea area or offshore area, the ship motion will be affected by the ocean and other disturbances. In (Kim et al., 2022), a Computational Fluid Dynamics (CFD) model was used to predict ship maneuverability in the currents, which can be perform very good performance in the simulation case, however it is difficult to validate in the real navigation with the ocean currents environment. To obtain the current information, a costly Doppler log is usually installed onboard. However, most of doppler logs can only obtain the ship velocity through water on longitudinal direction. Hence, it is necessary to find a common way to evaluate the ship velocity through water both on longitudinal and lateral direction.

8.2.1 Ship response model identification

The ship response model is a simplification of the ship hydrodynamic model and can be used for the study of ship heading control and short-term trajectory prediction [16]. It mainly includes first-order linear and nonlinear models, and second-order linear and nonlinear models. Owing to the complex forces and torques acting on the sailing ship, the motion relationship described by nonlinear models is usually more accurate to the actual ship motion than linear ones [17]. In this chapter, the first-order nonlinear response model is chosen for the description of ship motion:

$$T_N \dot{r} + r + \alpha r^3 = K_N(\delta + \delta_r), \tag{8.1}$$

where δ is the rudder angle, r is the yaw rate, and K_N, T_N are the dimensionless coefficients following the simplification of the hydrodynamic model, and α is the non-linear term coefficient, and δ_r is the rudder angle order. The values of the coefficients in the model are usually obtained by parameter identification using ship Zigzag-test data [18]. In this chapter, the least squares identification method is used, which is to find the optimal matching parameters for the minimum sum of error squares. A variation of Eq. 8.1 yields the least squares standard equation at time t.

$$\dot{r}(t) = \frac{1}{T_N}(t) - \frac{\alpha}{T_N}r(t)^3 + \frac{K_N}{T_N}\delta(t) + \frac{K_N \delta_r}{T_N}, \tag{8.2}$$

where $\dot{r}(t)$ is the observed variable, and $\{\frac{1}{T_N}, -\frac{\alpha}{T_N}, \frac{K_N}{T_N}, \frac{K_N \delta_r}{T_N}\}$ are the weight parameters to be determined, and $r(t)$, $r(t)^3$, $\delta(t)$ are functions on the states and control input.

According to the model Eq. 8.2, we assume that from $t = 1$, the model input and output data are sampled, with $y = \dot{r}$, $\boldsymbol{x}(t) = [r(t), r(t)^3, \delta(t), 1]$. Moreover, taking the length of the collected samples as N, the output data and information data $\{x(1), x(2), ..., x(N)\}$ are obtained. The extended matrix equation can be described as follows:

$$\boldsymbol{Y} = \boldsymbol{x}^T \boldsymbol{\omega},$$
$$\boldsymbol{Y} = [y(1), y(2), ..., y(N)]^T, \tag{8.3}$$
$$\boldsymbol{X} = [\boldsymbol{x}(1), \boldsymbol{x}(2), ..., \boldsymbol{x}(N)].$$

Here,

$$\boldsymbol{\omega} = \left[\frac{1}{T_N}, -\frac{\alpha}{T_N}, \frac{K_N}{T_N}, \frac{K_N \delta_r}{T_N}\right]^T.$$

The least squares problem can be solved by minimizing the loss function J to achieve the smallest $\hat{\boldsymbol{\omega}}$ to obtain the optimal solution. $\hat{\boldsymbol{\omega}}$ is the fitted value of $\boldsymbol{\omega}$, and J is expressed as follows:

$$J = \frac{1}{N}\sum_{t=1}^{N}\epsilon(t)^2 = \frac{1}{N}\sum_{t=1}^{N}\left[y(t) - \boldsymbol{X}^T(t)\boldsymbol{\omega}\right]. \tag{8.4}$$

It has a minimal value at the point where the derivative of $\boldsymbol{\omega}$ is 0, which is:

$$\frac{1}{N}[-2\boldsymbol{XY} + 2\boldsymbol{XX}^T\hat{\boldsymbol{\omega}}] = 0. \tag{8.5}$$

The least squares optimal solution is obtained by solving Eq. 8.5, which is:

$$\hat{\boldsymbol{\omega}} = \left[\boldsymbol{XX}^T\right]^{-1}\boldsymbol{XY}. \tag{8.6}$$

8.2.2 Extreme short-term prediction model for ships considering ocean currents

The three-Degree-Of-Freedom (3 DOF) kinematic model of the ship during its voyage is shown in Fig. 8.1. In this figure, the $x_n o_n y_n$ is the geodetic coordinate system, (x, y) is the coordinates of the ship in geodetic coordinates, θ is the angle between the ship course and due north direction, ϕ is the heading angle of the ship, β_s is the angle between the ship heading and direction of the velocity relative to water, V_s is the ship velocity through the water, and V is the ship velocity over ground.

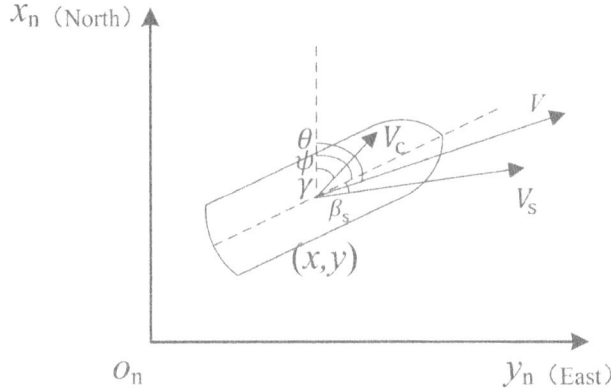

Figure 8.1: 3 DOF kinematic model of the ship.

The ship motion trajectory consists of the coordinate points at each moment, and the position of the coordinate points at future moments is determined by the ship velocity over ground, which is expressed as follows:

$$\begin{cases} \dot{x} = V_x \\ \dot{y} = V_y \end{cases}, \tag{8.7}$$

where V_x and V_y are the velocity over ground in the due north and due east direction, respectively. Their relationship is as follows:

$$\begin{cases} V_x = V \cos \theta \\ V_y = V \sin \theta \end{cases}, \tag{8.8}$$

The ship velocity over ground is the resultant one of the ship velocity through water and current velocity, where velocities in the direction of due north and due east are:

$$\begin{cases} V_x = V_{cx} + V_{sx} \\ V_y = V_{cy} + V_{sy} \end{cases}, \tag{8.9}$$

where V_{sx}, V_{sy}, V_{cx}, and V_{cy} are the components of the ship velocity through water, i.e., V_s, in the due north and due east direction, and the components of the sea current velocity, i.e., V_c, in the due north and due east direction, respectively. Their relationship is as follows:

$$
\begin{cases}
V_{sx} = V_s + \cos(\psi + \beta_s) \\
V_{sy} = V_s + \sin(\psi + \beta_s) \\
V_{cx} = V_c \cos \gamma \\
V_{cy} = V_c \sin \gamma
\end{cases},
\tag{8.10}
$$

In addition, the heading ψ with yaw rate r is denoted as: $\dot{\psi} = r$.

This chapter focuses on building an extreme short-term prediction of a ship trajectory in a uniform ocean current field, assuming the velocity V_c and direction γ of the current do not change in the prediction period. When a ship sails in a uniform ocean current field, its force characteristics are similar to those of sailing in calm water. Therefore, the angle between the ship course through water and ship heading are generated by the hydrodynamic influence when the ship moves relative to the current, the magnitude of β_s is usually only related to the rudder angle. Accordingly, considering that the ship rudder angle remains unchanged or changes slightly, β_s can be regarded as a constant value during the predicting period. With a short duration of the extreme short-term prediction time and slow changing dynamics of the ship velocity through water, V_s can be also considered as a constant value.

Set the state quantity $x = [x \ y \ V_x \ V_y \ \psi \ r]^T$, model input $u = \delta$, model output $y = [x \ y]^T$. By combining Eqs. 8.2 and 8.7-8.10, a model for extreme short-term prediction of ship trajectory is built as follows:

$$
\dot{x} = \begin{bmatrix}
x_3 \\
x_4 \\
-V_s x_6 \sin(x_5 + \beta_s) \\
V_s x_6 \cos(x_5 + \beta_s) \\
x_6 \\
-\dfrac{1}{T_N} x_6 - \dfrac{a}{T_N} x_6^3 + \dfrac{K_N}{T_N} u + \dfrac{K_N \delta_r}{T_N}
\end{bmatrix},
\tag{8.11}
$$

where $A = [1\ 1\ 0\ 0\ 0\ 0]$.

For the extreme short-term prediction, the constant values of V_s and β_s during the prediction period can be obtained by sensors, observers, or empirical data. Assuming that an extreme short-term prediction is made at time t, V_s, and β_s can be calculated by combining Eqs. 8.8-8.10.

$$
\begin{aligned}
V_s &= \sqrt{(V(t)\cos\theta(t) - V_c(t)\cos\gamma)^2 + (V(t)\sin\theta(t) - V_c(t)\sin\gamma)^2}, \\
\beta_s &= \arctan((V(t)\sin\theta(t) - V_c(t)\sin\gamma)/(V(t)\cos\theta(t) - V_c(t)\cos\gamma)) - \psi(t),
\end{aligned}
\tag{8.12}
$$

where the sea current velocity information, i.e., V_c and γ, can be obtained from sensors at time t. The algorithm of extracting the sea current velocity information will be described in the next section.

8.3 ALGORITHM OF EXTRACTING SEA CURRENT INFORMATION FROM TURNING TESTS

In actual navigation, ships are limited to the configuration of sensors on account of cost and practicability, which means the sea current velocity information cannot be obtained directly. To this end, this chapter proposes a sea current information extracted algorithm that can be applied in actual navigation. The Velocity Distribution Extraction (VDE) and Center Point Restoration (CPR) algorithm are proposed to extract the initial sea current information with turning test data, and the combination of the VDE and CPR algorithms is proposed to achieve more accurate prediction result.

8.3.1 Velocity distribution extraction (VDE) algorithm

When the ship turning test is conducted in calm water, after reaching a steady state, the ship will perform an approximately uniform circular motion [20], and the magnitude of the ship velocity over ground will remain constant. However, when a ship performs the turning test in the navigation environment of a uniform water flow field, its velocity over ground would be affected by the sea current velocity and direction, as shown in Fig. 8.2. When the ship turns to point A, the ship velocity over ground is maximum, and at this time the heading is θ_1, and the velocity over ground is denoted as V_{max}. When the ship turns to point B, the ship ground velocity is the smallest, at this time the heading is θ_2, and the velocity over ground is V_{min}. Owing to the uncertainty in the external environments, the difference between θ_1 and θ_2 is not exactly equal to 180°. Hence, θ_1 cannot be directly used as the direction of the sea current.

(1) If $\theta_1 > \theta_2$, the direction of the sea current γ is calculated as follows:

$$\gamma = 0.5(\theta_1 + \theta_2) + 90°. \tag{8.13}$$

(2) If $\theta_1 \leq \theta_2$, the direction of the sea current γ is calculated as follows:

$$\gamma = 0.5(\theta_1 + \theta_2) - 90°. \tag{8.14}$$

Thus, the velocity of the sea current V_y can be stated as:

$$V_y = 0.5\cos(\theta_1 - \gamma)(V_{max} - V_{min}). \tag{8.15}$$

By analyzing the ship velocity over ground during the ship turning test, the point of the maximum and the minimum velocity over ground can be obtained, and the sea current velocity can be calculated, the process of which is shown in Fig. 8.3.

8.3.2 Weighted derivation algorithm

When a ship is in the calm water for the turning experiment, the ship slewing center can be assumed as a fixed point. However, in the actual navigation scenarios, under the influence of the sea current, the ship track trajectory follows a spiral shape, and the ship slewing center point moves with the direction of the current. Therefore, the

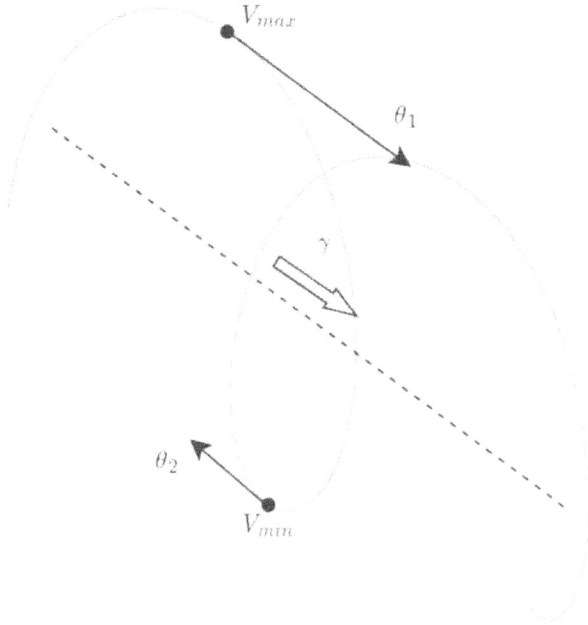

Figure 8.2: Maximum and minimum velocity illustration during the turning test.

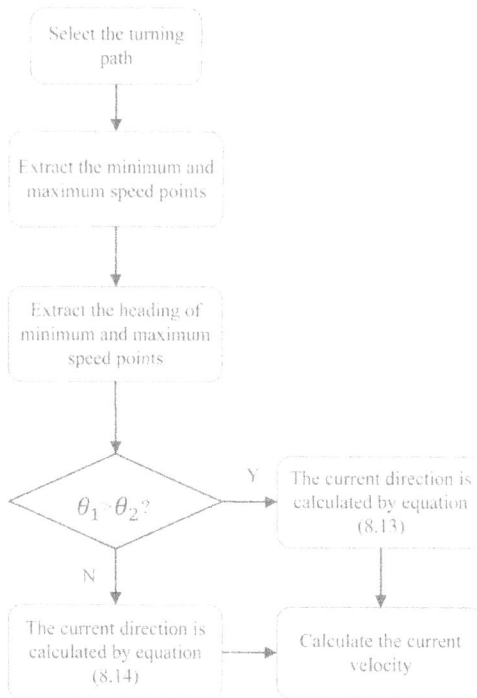

Figure 8.3: Velocity distribution and extraction of sea current.

connecting line of the ship slewing center points can be seen as the current direction, and the slewing point moving velocity can be seen as the current velocity, as shown in Fig. 8.4.

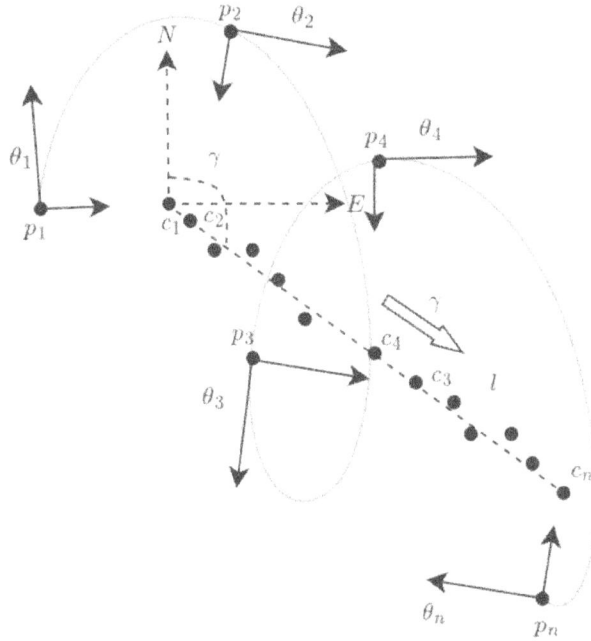

Figure 8.4: Slewing center point restoration during the turning test.

In Fig. 8.4, the trajectory point in the middle between the start and the end of the turning trajectory is $\{p_1, p_2, ..., p_n\}$, which corresponds to the heading direction $\{\theta_1, \theta_2, ..., \theta_n\}$.

(1) When the ship is turned clockwise, the ship slewing center coordinates is calculated as follows:

$$\begin{cases} x_c = x_p + R\cos(\psi + 90°) \\ y_c = y_p + R\sin(\psi + 90°) \end{cases}.$$ (8.16)

(2) When the ship is turned counterclockwise, the ship slewing center coordinates (x_c, y_c) can be calculated as follows:

$$\begin{cases} x_c = x_p + R\cos(\psi - 90°) \\ y_c = y_p + R\sin(\psi - 90°) \end{cases},$$ (8.17)

where R is the turning radius the same as one in calm water.

For the trajectory point $\{p_1, p_2, ..., p_n\}$, the coordinates of the trajectory points can be calculated as $\{c_1, c_2, ..., c_n\}$. The fitting of the ship trajectory points yields the line connecting the center point l, which starts at point c_1 and ends at point c_n. The direction of the line corresponds to the present direction of the ocean current in the region. The fitting of the distance between the line l with the starting point c_1

and the end point c_n yields the distance of current flow forward d in t_l. At this point, the current velocity v_c is:

$$v_c = d_l/(t_l), \qquad (8.18)$$

where t_l is the total time of the chosen turning test data.

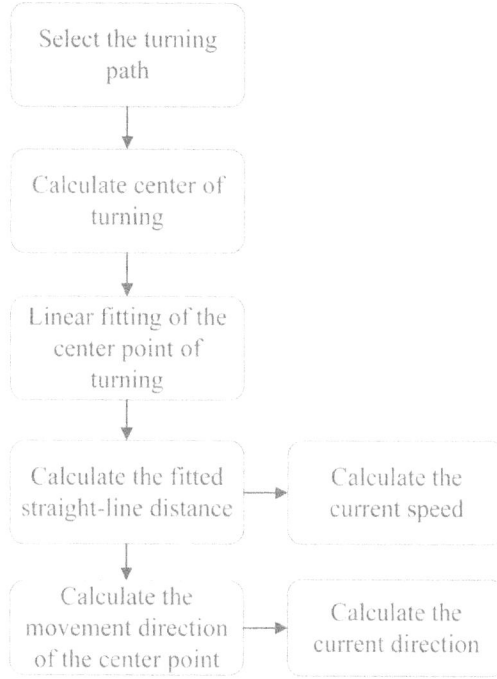

Figure 8.5: Flowchart of sea current derivation by CPR.

In summary, according to the ship actual turning trajectory and heading, the ship slewing center can be derived based on the ship turning radius in the calm water and the ship heading at the present moment. The average current velocity in the turning area can be obtained by fitting the ship slewing center point and calculating the distance covered by the ship slewing center point during the turning time. The current direction in the test area can be derived from the direction of the fitted straight line. Fig. 8.5 depicts a flow chart summarizing the CPR method.

8.4 EXTREME SHORT-TERM PREDICTION METHOD CONSIDERING OCEAN CURRENT

The deviation of the maximum velocity over ground from the predicted current direction with the ship VDE method is attributed to uncertainties in the actual navigation, which is denoted as e_1. Similarly, the mean error of the straight-line trajectory with CPR is denoted as e_2. Assuming that the velocity and direction of the sea current obtained by the VDE method are V_e and θ_e, while the velocity and direction of the sea current obtained in the CPR method are V_m and θ_m. Subsequently, the weighted derived current velocity Vc and direction θ_c direction are:

$$\begin{cases} V_c = \dfrac{e_2}{e_1 + e_2} V_e + \dfrac{e_2}{e_1 + e_2} V_m \\ \theta_c = \dfrac{e_2}{e_1 + e_2} \theta_e + \dfrac{e_2}{e_1 + e_2} \theta_m \end{cases}. \tag{8.19}$$

Moreover, considering VDE or CPR could lead to relatively large errors, the acceptable range for the deviations of VDE and CPR is accordingly set as e_1 max and e_2 max . When the deviation value of VDE or CPR is larger than its acceptable range, VDE or CPR method is not considered for the prediction of current velocity and direction. At this time, V_c can be expressed as follows:

$$\begin{cases} V_c = V_e, \theta = \theta_c, if\ e_1 \le e_{1\ max}, e_2 > e_{2\ max} \\ \theta_c = V_m, \theta = \theta_m, if\ e_2 \le e_{2\ max}, e_1 > e_{1\ max} \end{cases}. \tag{8.20}$$

Note that if the deviations of both methods are both greater than their acceptable range, it is recommended that the turning test needs to be redone.

The process of the sea current weighted derivation method is shown in Fig. 8.6.

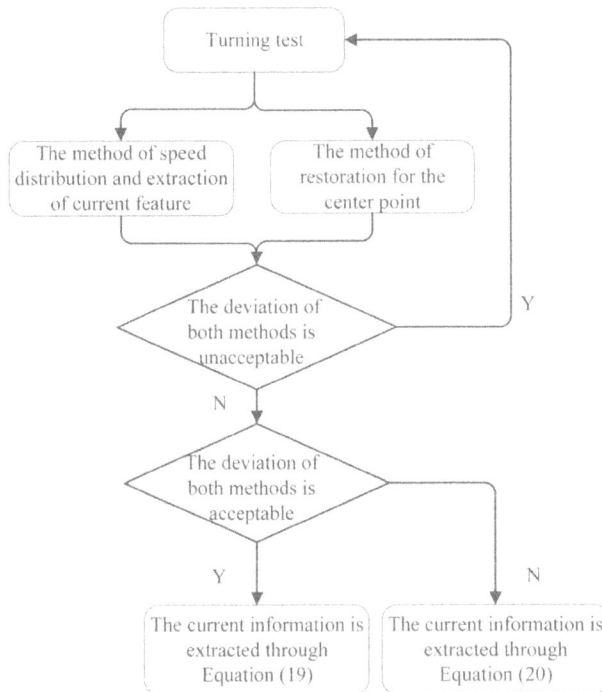

Figure 8.6: Flow chart for sea current weighted derivation process.

8.4.1 Extreme short-term prediction for ship trajectories

This chapter combines the response model and sea current information derivation method obtained from the identification, and the ship extreme short-term prediction model, to predict the ship trajectory in the extreme short-term (usually short than 20 s). Moreover, considering that the real-time sea current information would deviate from the original extracted sea current information owing to the continuous movement of the ship navigation geographic location during navigation, the predicted sea current information is online corrected through the observation of the prediction errors to achieve more accurate trajectory prediction.

8.4.2 Extreme short-term prediction of ship trajectory based on the online correction

Ship trajectory prediction can help improve the accuracy of the ship trajectory control [16]. It can predict the changes in the ship trajectory in the next moments with different control inputs. For instance, when a ship is tracking along the excepted trajectory, the proposed extreme short-term prediction model can be used to optimize, the solution and obtain the excepted control inputs. However, owing to the large inertia and time delay, the prediction of the ship trajectories in multiple moments is required for more accurate and stable control.

Based on the response model obtained from the identification and extracted sea current information, the ship motion trajectory in the extreme short term can be predicted based on the ship extreme short term prediction model, the principle of which shown in Fig. 8.7.

Defining the extreme short-term prediction step size is i, in combination with Eq. 8.12 and the defined ship response model parameters, V_s and β_s, the states in future times can be predicted. Considering that the deviations between predictive states and actual states become larger when the predictive time duration goes longer, the predictive states should be updated based on the latest measured states with an appropriate control period. For the ship trajectory control, the short-term prediction period is set as 15 s in this paper.

8.5 EXPERIMENTAL VALIDATION AND ANALYSIS

In the actual navigation trajectory prediction, owing to the continuity of the ship trajectory geographic location, differences would exist between the real-time sea current information and the original derived current information [19]. Accordingly, predictive errors will occur when the prediction is only based on the original current information. The error is mainly generated as Eq. 8.12 and caused by the deviations in the sea current information V_c and γ.

Considering that the prediction interval is short, the current changes slowly, and the relationship between the error and the predicted current can be obtained by the extreme short-term prediction model, the realtime prediction error can be used to correct the current information online and update the real-time current information to obtain more accurate predictive results. In ignorance of the influence of uncertainties, the relationship between the sea current observation errors and the trajectory

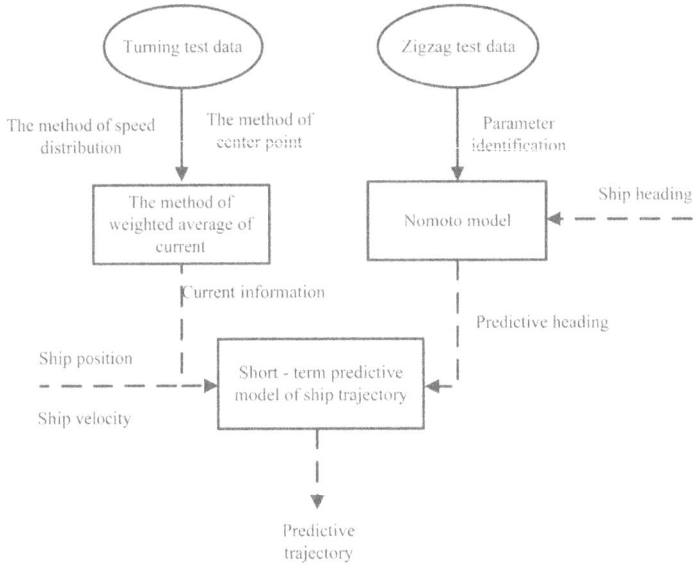

Figure 8.7: Principle of ship trajectory prediction.

predictive errors can be derived from Eq. 8.11 and 8.12:

$$
\begin{cases}
V_{ox}(t) = \dfrac{V_{errx}(t)}{2(1+\cos(\psi(t)-\psi(t-i)))} - \dfrac{V_{erry}(t)}{2\sin(\psi(t)-\psi(t-i))} \\
V_{oy}(t) = \dfrac{(1+\cos(\psi(t)-\psi(t-i)))V_{erry}(t)}{2\sin^2((\psi(t)-\psi(t-i)))} + \dfrac{V_{errx}(t)}{2\sin(\psi(t)-\psi(t-i))}
\end{cases}, \tag{8.21}
$$

where $V_{ox}(t)$ is the sea current velocity in the x direction at the time t, $V_{oy}(t)$ is the velocity correction deviation in the y direction of the current at time t, $\psi(t)$ is the ship heading at time t, $\psi(t-i)$ is the ship heading at the time $t-i$ with i being the prediction interval step, $V_{errx}(t)$ is the deviation in the x direction of the predicted velocity from the actual velocity at the time t, and $V_{erry}(t)$ is the deviation in the y direction of the predicted velocity from the actual velocity at the time t. The deviation from the direction can be obtained from the trajectory prediction error, which is as:

$$
\begin{cases}
V_{errx}(t) = \dfrac{x(t)-x_p(t)}{iT_s} \\
V_{erry}(t) = \dfrac{y(t)-y_p(t)}{iT_s}
\end{cases}, \tag{8.22}
$$

where $x_p(t)$ and $y_p(t)$ are the coordinates of the trajectory point at the time t, $x(t)$ and $y(t)$ are the coordinates of the actual trajectory points at the time t, and T_s is the length of time between each moment.

Once the corrections have been obtained, the corrected sea current information can be calculated.

$$\begin{cases} V_c'(t) = \sqrt{(V_c(t)\cos\gamma(t) + V_{ox}(t))^2 + (V_c(t)\sin\gamma(t) + V_{oy}(t))^2} \\ \gamma'(t) = \arctan\dfrac{V_c(t)\sin\gamma(t) + V_{oy}(t)}{V_c(t)\cos\gamma(t) + V_{ox}(t)} \end{cases} \qquad (8.23)$$

In the predicting process, the original sea current information is replaced by the corrected and real-time sea current information, to achieve more accurate extreme short-term prediction of ship trajectory. The prediction process is shown in Fig. 8.8.

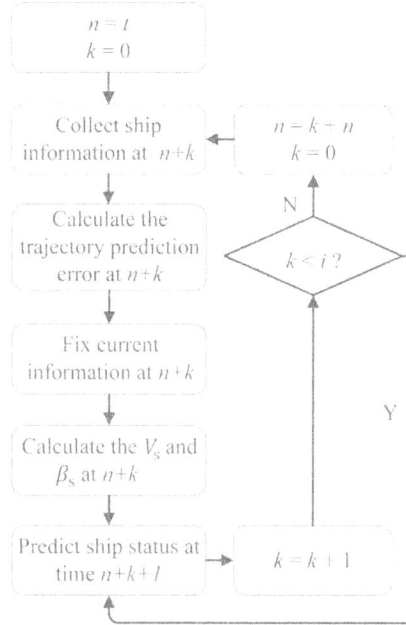

Figure 8.8: Flow chart for the extreme short-term ship trajectory prediction based on online corrections.

In this chapter, the test verification is conducted in the sea area of Zhoushan, China. The sea area belongs to sheltered sea areas, where the influence of wind and waves on the ship motion trajectory is small and the ship motion is mainly disturbed by the sea current. These tests were conducted on May 27, 2022. The test ship is a trimaran, and its relevant parameters and propulsion working conditions are listed in Tables 8.1 and 8.2, respectively. Working condition two is denoted as the stable cruising condition, which is mainly employed for test verification. The test scenarios are shown in Fig. 8.9.

It should be noted that considering the own ship navigation could be interfered by surrounding ships, static obstacles like islands or reefs in the test area, the manipulation of steer is difficult to keep constant according to the expected steps during the experiment. In the case of this chapter, the own ship encountered a collision risk with another ship during the Zigzag-test process, as shown in Fig. 8.10 (a), and therefore, a collision avoidance operation was conducted (Fig. 8.10 (b), 60 s–80 s). However, it

TABLE 8.1   Parameters of the trimaran ship.

Parameter	Value	Parameter	Value
Length	45.67 m	Main waterline length	44.65 m
Length between perpendiculars	42.97 m	Main waterline breadth	5.30 m
Molded breadth	11.92 m	Main displacement	169.6 t
Molded depth	4.00 m	Demihull center distance	11.00
Design draft	1.60 m	Demihull waterline length	14.72 m
Design discharge	174.2 t	Demihull molded depth	3.00 m
Designed velocity	12 knots	Demihull waterline breadth	0.59 m
Main molded breadth	6.00 m	Individual demihull discharge	2.3 t

TABLE 8.2   Ship propulsion working conditions.

Working condition	Rated rotation speed
No. 1	80 rpm
No. 2	180 rmp
No. 3	280 rmp
No. 4	390 rmp
No. 5	430 rmp

Figure 8.9: The test scenario in Zhoushan

does not affect the identification results since the heading also changes accordingly with the rudder angle based on the model. Therefore the data during this test process can be used for the parameter identification.

Parameter identification of the Zigzag-test data is conducted using the method proposed in this chapter. The identification parameters and results are shown in Table 8.3. It is worth noting that in this chapter, points 50–270 of the navigation data are used for the parameter identification. The ship states in step 1 (the first 50 steps reserved for the validation set) are taken as the inputs to predict the states of step 2,

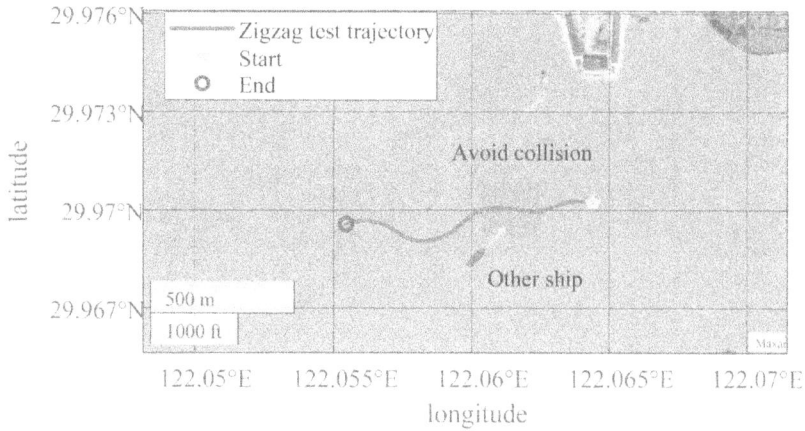

(a) Actual trajectory during the Zigzag test.

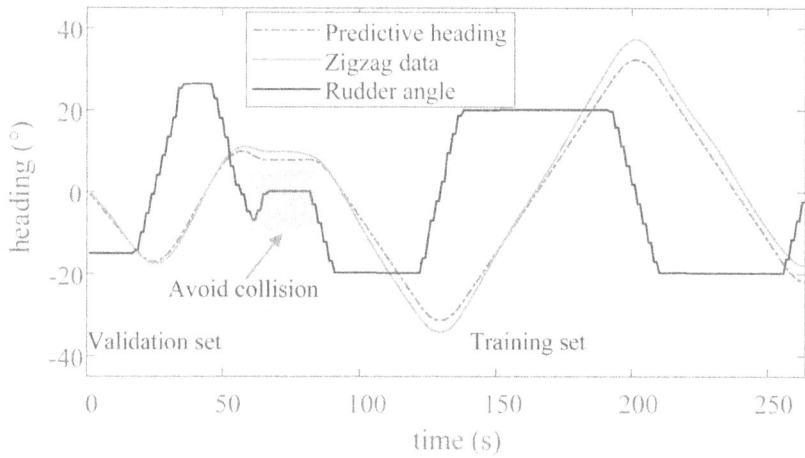

(b) Ship heading and rudder angle.

Figure 8.10: Ship Zigzag-test in sea with currents.

TABLE 8.3 Ship model identification parameters and results.

Parameter	Value
Sampling period (Hz)	1
Length of training set	220
Length of validation set	50
Iteration number of the least square method	219
K	0.06
T	0.65
a	0.30
δ_r	0.17

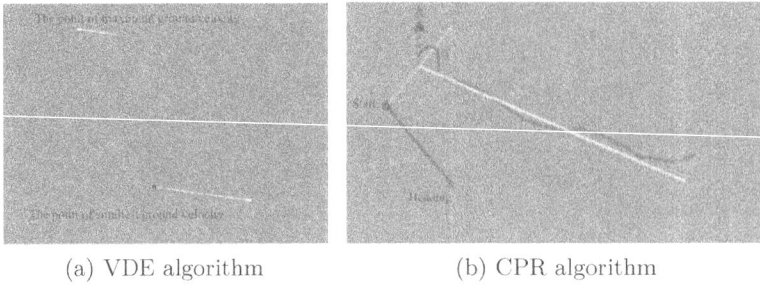

(a) VDE algorithm   (b) CPR algorithm

Figure 8.11: Sea current derivation test with different algorithms.

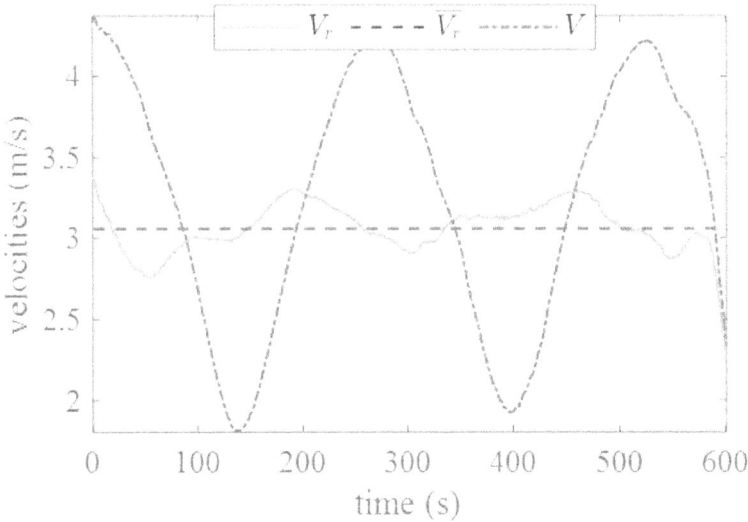

Figure 8.12: Ship velocity distribution during the turning test.

and the predicted value is put into the model to iterate the states of the future step n. The model short-term forecasting results (1–50 s) are satisfied, but the accuracy decreases over time due to accumulated errors.

After obtaining the ship response model, we performed current derivation by the left turning test. Fig. 8.11 (a) shows the sea current derivation by the VDE method, and Fig. 8.11 (b) shows the sea current derivation by the CPR method.

During the left turning test, the maximum ground velocity is 4.2 m/s, when the heading of the ship is 97.18°. The minimum velocity over ground is 1.8 m/s, when the heading of the ship is 277.03°.

1) Based on the VDE method, the current flow direction is 97.2° and velocity is 1.2 m/s. When the main engine speed of the ship is 180 rpm, the left turning radius of the ship in calm water is 134 m.

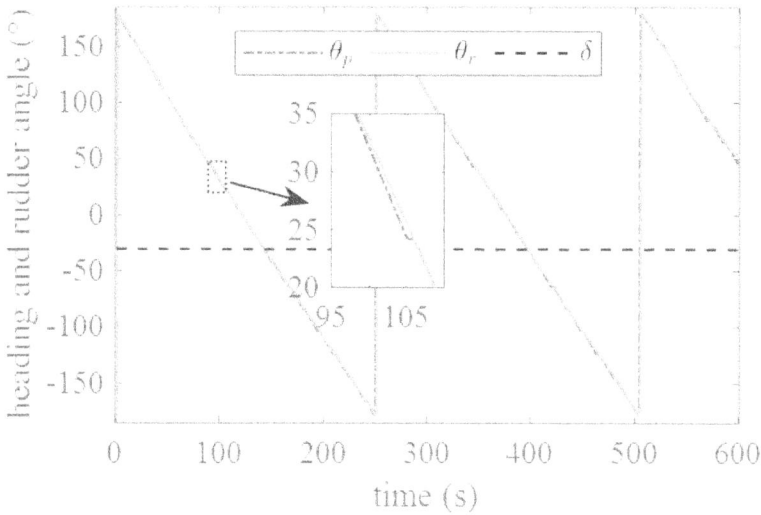

Figure 8.13: Ship heading prediction with currents.

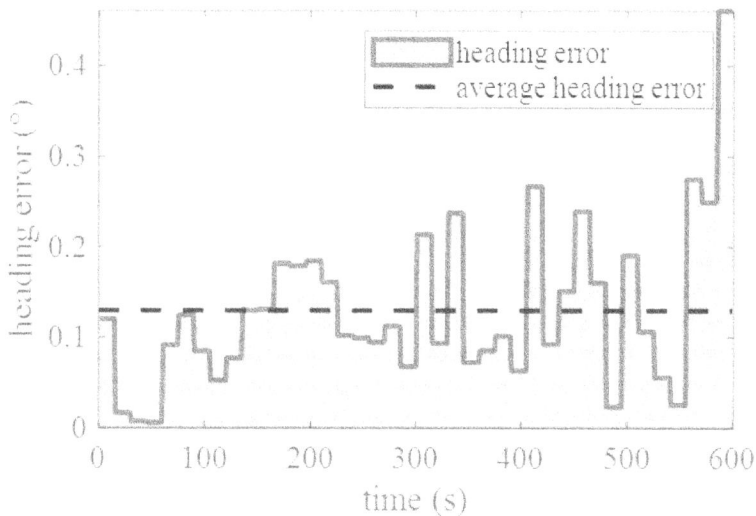

Figure 8.14: Prediction error for the ship heading.

2) Based on the CPR method, the fitted line segment of the ship slewing center point could be calculated as the angle between the ship fitted straight line and due north, i.e., 97.0°. Moreover, the ship slewing center point moved 671 m in the flow area, through 591 s under a current velocity of 1.14 m/s.

The difference between the current velocity and direction calculated by the VDE and CPR method is small from the calculation results. In this scenario, either VDE or CPR can be chosen. The derived sea current information is set as a velocity of 1.2 m/s and direction of 97.2°.

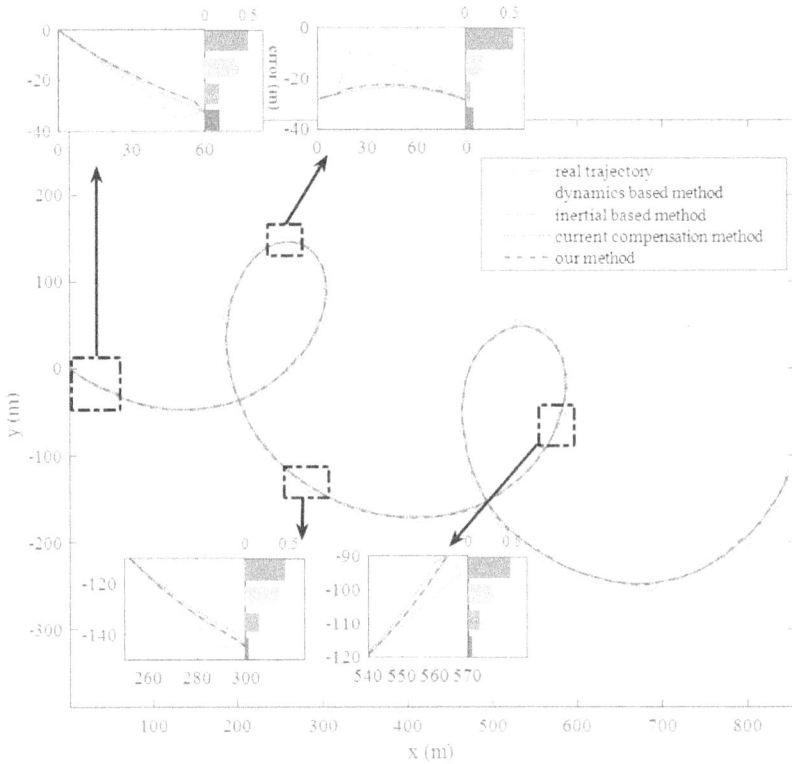

Figure 8.15: Results for the ship trajectory prediction.

After obtaining the current data and ship model parameters, the ship heading prediction test needs to be carried out. The ship is set to carry out the left turning experiment in working condition 2, the rudder order is kept at $30°$ left rudder, and the length of each prediction is 15 s. A total of 40 sets of ship state parameters are predicted, and the sampling frequency of the ship heading, rudder order, heading angle and other information is 1 Hz. The starting point of the ship heading prediction is $(29.971912°\text{N}, 122.053981°\text{E})$, and the prediction results are shown in Fig. 8.13 – Fig. 8.16. In Fig. 8.12, V_r, $\bar{V}_r$, and V are the ship velocity through water, the ship average velocity through water, and the ship velocity over ground, respectively. In Fig. 8.13, θ_p, θ, and δ are the predicted ship heading, actual heading, and rudder angle commands, respectively.

Fig. 8.13 shows the ship velocity through water approximated by bringing the derived current velocity and direction into the actual velocity over ground. Subsequently, the excepted velocity through water is nearly the same as the average velocity in the turning experiment under working condition 2. The ship heading prediction is shown in Fig. 8.13, and the errors of each segment of prediction are shown in Fig. 8.14, where the maximum prediction error is $0.4610°$ (occurred after the ship decelerated, hence the error became larger here) and the average prediction error is $0.1299°$.

The proposed method is compared with the dynamics-based method, inertia-based method, and current compensation method based on current data correction

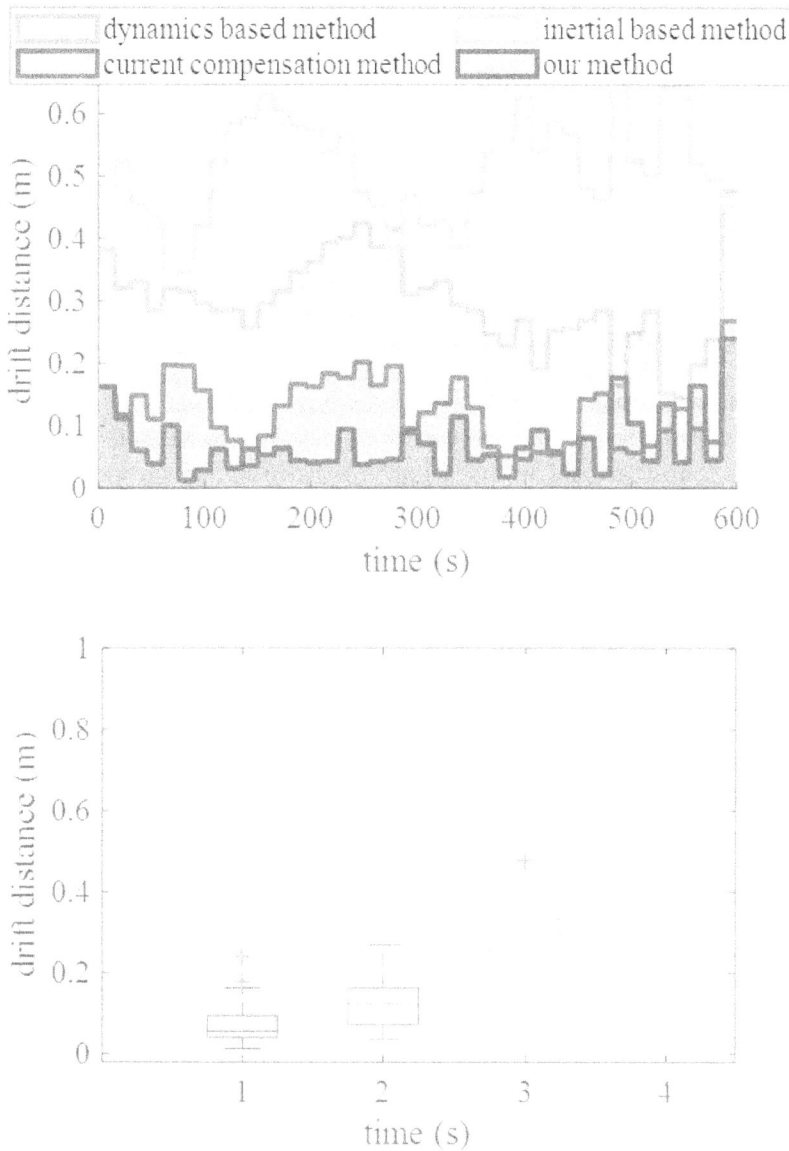

Figure 8.16: Errors for the ship trajectory prediction with different methods.

prediction. The dynamics-based method is a prediction method that does not consider the effect of sea currents in the prediction process. The inertia-based prediction method is a prediction method that does not consider the ship dynamics model. It assumes that the ground velocity will not change during the prediction period and that the magnitude and direction of the velocity over ground will remain the same as the present value. The method based on current data correction prediction refers to a prediction method that considers the influence of sea current with dynamical constraints, but a Markov chain is not considered in the prediction process to correct the prediction results. The prediction comparison results are shown in Figs. 8.15 and 8.16, which show the segmented prediction trajectory map and trajectory prediction error map, respectively. The results show that the median error of the proposed method is 0.056 m (the average value within 15 s, when the prediction time duration is i periods, assuming the endpoint of the predicted trajectory is P_1 and actual endpoint of this section of the trajectory is P_2, the average error at the time could be regarded as $||P_2 - P_1||_2/i$), where the differences in errors compared to the dynamic-based method, and current compensation method are reduced by 80.6%, 89.3%, and 54.7%, respectively.

8.6 CONCLUSIONS

In this chapter, we propose VDE, CPR, and weighted derivation method to extract the velocity and direction of the sea current. Accordingly, we consider the influence of ship motion characteristics and sea current disturbances on the accuracy of the trajectory prediction. A trajectory prediction model based on the ship response model and 3 DOF kinematic model is also developed. Moreover, a segmented extreme short-term ship trajectory prediction method is proposed based on the ship trajectory prediction model and the predicted sea current, while the current information is corrected online during the prediction process based on the prediction errors. Finally, the proposed prediction method is validated and analyzed using the turning test data. The field test results show that the proposed extreme short-term trajectory prediction method can accurately predict the trajectory. The median error of the prediction within 15 s is 0.056 m. Compared with the prediction method based on ship dynamics, inertia, and constant sea current, the prediction error of the proposed method is decreased by 80.6% (dynamic-based method), 89.3% (inertial-based method) and 54.7% (current compensation method), respectively.

In future research, we will consider studying the method of extreme short-term ship trajectory prediction under a non-constant water flow field, wave force, or wind force to further improve the accuracy of ship trajectory prediction under environmental disturbances and ensure the safety of ship navigation. We will also consider establishing a 6 DOF Mathematical Model Group (MMG) model to achieve long-term prediction.

Bibliography

[1] D. Zhang, X. Chu, W. Wu, Z. He, Z. Wang Z. Liu C., "Model identification of ship turning maneuver and extreme short-term trajectory prediction under the influence of sea currents," *Ocean Engineering*, vol. 278, p. 114367, 2023.

[2] Z. He, C. Liu, X. Chu, R.R. Negenborn, Q. Wu, "Dynamic anti-collision A-star algorithm for multi-ship encounter situations," *Applied Ocean Research*, vol. 118, p. 102995, 2022.

[3] J. Wang, L. Zou, D. Wan, "Numerical simulations of zigzag maneuver of free running ship in waves by rans-overset grid method," *Ocean Engineering*, vol. 162, pp. 55–79, 2018.

[4] Q. Zhang, X.K. Zhang, N.K. Im, "Ship nonlinear-feedback course keeping algorithm based on mmg model driven by bipolar sigmoid function for berthing," *International Journal of Naval Architecture and Ocean Engineering*, vol. 9, no. 5, pp. 525–536, 2017.

[5] T.I. Fossen, "A nonlinear unified state-space model for ship maneuvering and control in a seaway," *International Journal of Bifurcation and Chaos*, vol. 15, no. 9, pp. 2717–2746, 2005.

[6] R. Skjetne, T.I. Fossen,, P.V. Kokotović, "Adaptive maneuvering, with experiments, for a model ship in a marine control laboratory," *Automatica*, vol. 41, no. 2, pp. 289–298, 2005.

[7] S. Li, C. Liu, X. Chu, M. Zheng, Z. Wang, J. Kan, "Ship maneuverability modeling and numerical prediction using CFD with body force propeller," *Ocean Engineering*, vol. 264, p. 112454, 2022.

[8] M. Ueno, Y. Tsukada, "Rudder effectiveness and speed correction for scale model ship testing," *Ocean Engineering*, vol. 109, pp. 495–506, 2015.

[9] M. Hasnan, H. Yasukawa, N. Hirata, D. Terada, A. Matsuda, "Study of ship turning in irregular waves," *Journal of Marine Science and Technology*, vol. 25, no. 4, pp. 1024–1043, 2020.

[10] H. Yasukawa, N. Hirata, A. Matsumoto, R. Kuroiwa, S. Mizokami, "Evaluations of wave-induced steady forces and turning motion of a full hull ship in waves," *Journal of Marine Science and Technology*, vol. 24, pp. 1–15, 2019.

[11] J.H. Lee, Y. Kim, "Study on steady flow approximation in turning simulation of ship in waves," *Ocean Engineering*, vol. 195, p. 106645, 2020.

[12] G. Li, X. Zhang, "Research on the influence of wind, waves, and tidal current on ship turning ability based on norrbin model," *Ocean Engineering*, vol. 259, p. 111875, 2022.

[13] J. Lou, H. Wang, J. Wang, Q. Cai, H. Yi, "Deep learning method for 3-dof motion prediction of unmanned surface vehicles based on real sea maneuverability test," *Ocean Engineering*, vol. 250, p. 111015, 2022.

[14] C.Y. Tzeng, J.F. Chen, "Fundamental properties of linear ship steering dynamic models," *Journal of Marine Science and Technology*, vol. 7, no. 2, 2009.

[15] Q. Sun, Z. Tang, J. Gao, G. Zhang, "Short-term ship motion attitude prediction based on LSTM and GPR," *Applied Ocean Research*, vol. 118, p. 102927, 2022.

[16] C. Liu, R.R. Negenborn, X. Chu, H. Zheng, "Predictive path following based on adaptive line-of-sight for underactuated autonomous surface vessels," *Journal of Marine Science and Technology*, vol. 23, pp. 483–494, 2018/

[17] K. Nomoto, K. Taguchi, K. Honda, S. Hirano, "On the steering qualities of ships," *Journal of Zosen Kiokai*, vol. 99, pp. 75–82, 1956.

[18] W. Wu, X. Chu, C. Liu, H. Zheng, Z. He, "Predictive longitudinal following control for ship platoon considering diesel engine driven propeller reversal," *Ocean Engineering*, vol. 263, p. 112231, 2022.

[19] G. Budak, S. Beji, "Controlled course-keeping simulations of a ship under external disturbances," *Ocean Engineering*, vol. 218, p. 108126, 2020.

[20] M. Ueno, Y. Yoshimura, Y. Tsukada, H. Miyazaki, "Circular motion tests and uncertainty analysis for ship maneuverability," *Journal of Marine Science and Technology*, vol. 14, pp. 469–484, 2009.

III

Applications over the Water

CHAPTER 9

Waterborne Applications of USVs

9.1 INTRODUCTION

As with the maturity of the USV NGC technologies and availability of affordable hardware components, USVs have seen increasingly wider applications in various occasions. The advantages of USVs compared to the traditional crewed ships have been elaborated in Chapter 1.

The applications of USVs, mostly with small sizes, first emerge in the field of military, and then in the fields of ocean science, mineral detection etc. Larger sizes of USVs are later seen for the autonomous waterborne transportation [1, 2]. In this chapter, we briefly discuss three major applications of USVs, i.e., in hostile military scenarios, scientific oceanographic observation, and intelligent waterborne transportation. It should be noted that the applications of USVs are definitely not confined to the mentioned three cases. However, different applications rely on mostly the same core capabilities of USVs that are introduced in the earlier chapters of the book.

9.2 MILITARY SCENARIOS

USVs have become an important component in modern military operations, playing a significant role in enhancing combat efficiency, reducing personnel risks and costs [3]. Due to the mobility and flexibility, USVs are widely used in surveillance, reconnaissance and patrolling tasks [4]. For example, Zyvex Technologies developed the Piranha USV for reconnaissance [5]. Norway developed the Seastar USV for strategy, patrols, reconnaissance, interference and decoy, and extended optical magnetic field [6]. The USV could also be remote weapons platform with attack capabilities by deploying weapons on the USV, i.e., the suicide USVs. Besides, USVs can be used in force protection missions [7].

German Navy has used an M8 series of USVs as fast attack crafts for anti-terrorism tasks [5]. The Protector USVs of the Israeli company Rafael have been known as the outstanding performance in reconnaissance, counter-mine warfare, and electronic warfare [8]. The Protector USVs perform so well that the Singapore Navy have bought several Protector USVs to conduct a wide spectrum of critical missions, without exposing personnel and capital assets to unnecessary risks.

Due to the cross-medium property, USVs can also effectively respond to threats from submarines, and thus are an important part of the anti-submarine warfare [9]. US Navy, French navy and Singapore navy have also adopted the Spartan USVs for reconnaissance, counter-mine and antisubmarine applications [3]. As the development of USV technology progresses, it is expected that the USV military applications could become increasingly diverse and expansive.

9.3   OCEANOGRAPHIC OBSERVATION

The oceanographic observation practices have been revolutionized with the unmanned marine platforms, including both the USVs and autonomous underwater vehicles (AUVs). Traditionally, the oceanographic observation has been dominated by costly manned vessels, the measurement scales of which are in general limited due to logistic and technological reasons. For oceanographic observations, USVs are usually used as the platform to carry scientific payloads to collect the oceanographic information along the USV trajectory [10], showing paramount advantages in terms of cost, safety, mobility, and efficiency.

There are mainly two types of oceanographic applications with USVs. The first type involves measuring the biological or chemical parameters [11] of the water volume with the sensors on USVs. Due to the relatively easy accessibility to the communication and localization resources on the surface, USVs have been coordinated with AUVs to conduct Lagrangian studies and in-situ observations of microbial communities in the deep chlorophyll maximum layer [12]. The USV is equipped with satellite communications and underwater acoustic tracking capabilities. While acoustically tracking the AUV platforms performing the sampling missions, the USV also acts as a communication relay, connecting the AUV with the operator to provide situational awareness and intervene if needed.

Another interesting application of USVs can be found in [13] where USVs are used to observe and quantify the behavior and sensitivity to light pollution of the zooplankton in the high Arctic polar night. Accordingly, the USVs need to be equipped with the hyperspectral irradiance sensor and acoustic profiler. The C-Worker 5 work-class USV [14] has adaptive and autonomous water quality profiling capability to perform water column sampling. In addition, the USV can transmit hypoxia monitoring data in real time to support the investigation and modeling of the hypoxia zone in the northern Gulf of Mexico. With the efficient sampling of the marine environmental parameters along the USV trajectories, USVs have become useful tools for ocean biological and chemistry scientists.

To obtain the distribution of the light intensity across the water surface, a light intensity sensor has been installed on the USV platform as depicted in Fig. 6.1. Due to the mountain near the sea area, the sunlight is blocked resulting in significant variations in surface light intensity. The USV is first remotely controlled to travel through the light intensity field, capturing light intensity data at 2 Hz. Based on the observational data, Gaussian process regression is utilized to estimate the distribution

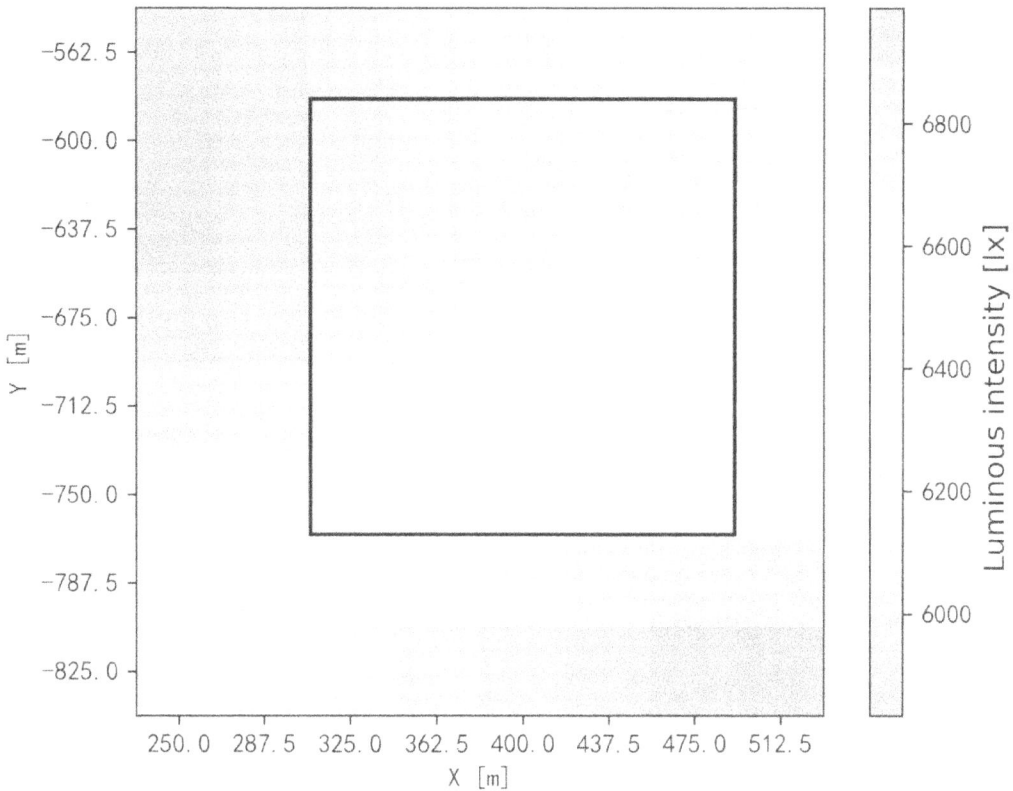

Figure 9.1: Estimated light intensity field with the sampled data with a USV.

of the light intensity across the water surface. The estimated light intensity field is illustrated in Fig. 9.1.

The second type of oceanographic applications involves ocean mapping and bathymetric survey with USVs [10][15][16]. The ocean mapping and bathymetric survey tasks have been tedious and expensive with the traditional manned motherships. However, by using USVs with geographic survey sensors on board, e.g., the multibeam acoustic sonar, the target sea area can be efficiently mapped with specifically designed path planning algorithms for the USV [16]. The non-standard USV with integrated multi-beam echo sounders has been used for the acoustic seafloor mapping [17]. The USV system has the advantages of light weight, low load and shallow draft, making it suitable for operation in shallow water and areas that are difficult to reach with traditional hydrographic vessels. The USV is capable of collecting highly accurate bathymetric data and enabling high-resolution mapping of the seafloor.

9.4 INTELLIGENT WATERBORNE TRANSPORTATION

With the mobility, USVs can naturally be used as autonomous waterborne transportation vehicles, either for cargoes [18][2][1][19] or passengers [20].

The concept of waterborne AGVs has been proposed in [1] which can be seen as the counterparts of the conventional automated guided vehicles. Waterborne AGVs

Figure 9.2: ITT waterway network for USVs in the Maasvlakte area, the port of Rotterdam, from Google Earth [21].

are essentially a type of USVs that are dedicated to short-distance autonomous waterborne transportation, e.g., the inter terminal transportation (ITT). Advanced control theories have been proposed for both a single and multiple waterborne AGVs. However, the idea of autonomous ITT with waterborne AGVs has only been validated with comprehensive simulations.

Notably, for the autonomous dynamic waterborne ITT in, e.g., port areas in the port of Rotterdam in the Netherlands (Fig. 9.2) or the archipelago areas in Zhoushan, China, a group of USVs have been scheduled with a dynamic rolling horizon scheduling strategy. The problem is mathematically modeled in a rolling horizon fashion considering dynamically arriving waterborne transportation requests, current USV states, dynamic waterway transport network, time windows of transportation requests, USV capacity limits and load/unload service times at terminals. Simulation results based on realistic ITT dataset and waterway network in the port of Rotterdam have demonstrated the effectiveness of such an autonomous ITT system with USVs. The fully autonomous waterborne transportation with USVs shows great potential to revolutionize and benefit the shipping industry.

The Mariner USVs, developed by a Norwegian company, tested the fully autonomous short distance waterborne transportation with USVs in 2023 [18]. In China, the concept "New Generation of Waterborne Transportation" with USVs has been proposed [22]. Such new generation of intelligent waterborne transportation system is composed of autonomous ships, digitized waterways, smart ports, and intelligent management system. This new system has been applied in different scenarios, e.g.,

Yangtze River, Qinhuai River, etc. The "ship-based monitoring and watch-keeping" mode is thought as the trend of future vessel operations.

In Europe, the USV project, NOVIMAR, was launched in 2017. The concept of "vessel train" is proposed to improve the navigation efficiency and lower the labor cost [23]. Numerical experiments, scaled model ship experiments and full-size ship experiments have been carried out to validate the energy saving and efficiency of such vessel train systems. Similarly, DARPA of US launched the Sea Train program. The main purpose is to demonstrate long range deployment capabilities for a distributed fleet of tactical USVs [24]. As a typical civilian application of USVs, intelligent waterborne transportation could further benefit from the development USVs and also from cooperative USVs.

Bibliography

[1] H. Zheng, "Coordination of Waterborne AGVs," Ph.D. dissertation, Delft University of Technology, Delft, The Netherlands, 2016.

[2] T. K. Grose, "Skip the skipper," *ASEE Prism*, vol. 31, no. 5, pp. 8–8, 2022.

[3] V. Ricci, M. Allen, J. M. Monti, and A. L. Goodman, "SPARTAN: A distributed mobile USW sensor and weapon system," Naval Undersea Warfare Center DIV Newport RI, Tech. Rep., 2002.

[4] T. Pastore and V. Djapic, "Improving autonomy and control of autonomous surface vehicles in port protection and mine countermeasure scenarios," *Journal of Field Robotics*, vol. 27, no. 6, pp. 903–914, 2010.

[5] W.-R. Yang, C.-Y. Chen, C.-M. Hsu, C.-J. Tseng, and W.-C. Yang, "Multifunctional inshore survey platform with unmanned surface vehicles," *International Journal of Automation and Smart Technology*, vol. 1, no. 2, pp. 19–25, 2011.

[6] Z. Liu, Y. Zhang, X. Yu, and C. Yuan, "Unmanned surface vehicles: An overview of developments and challenges," *Annual Reviews in Control*, vol. 41, pp. 71–93, 2016.

[7] S. Campbell, W. Naeem, and G. W. Irwin, "A review on improving the autonomy of unmanned surface vehicles through intelligent collision avoidance manoeuvres," *Annual Reviews in Control*, vol. 36, no. 2, pp. 267–283, 2012.

[8] E. Simetti, A. Turetta, G. Casalino, E. Storti, and M. Cresta, "Protecting assets within a civilian harbour through the use of a team of USVs: Interception of possible menaces," in *Proceedings of IARP workshop on robots for risky interventions and environmental surveillance-maintenance*, 2010, pp. 1–6.

[9] J. Su, Y. Yao, and W. Wang, "Research on rules of engagement for large-scale unmanned surface vehicle in anti-submarine warfare," in *Proceedings of 2021 International Conference on Autonomous Unmanned Systems (ICAUS 2021)*. Singapore: Springer, 2022, pp. 197–206.

[10] K. Zwolak, B. Simpson, B. Anderson, E. Bazhenova, R. Falconer, T. Kearns, H. Minami, J. Roperez, A. Rosedee, H. Sade, N. Tinmouth, R. Wigley, and Y. Zarayskaya, "An unmanned seafloor mapping system: The concept of an AUV integrated with the newly designed USV SEA-KIT," in *In Proceedings of OCEANS*, Aberdeen, UK, 2017, pp. 1–6.

[11] A. Guerrero-González, F. García-Córdova, and F. Ortiz, "A multirobot platform based on autonomous surface and underwater vehicles with bio-inspired neuro-controllers for long-term oil spills monitoring," *Autonomous Robots*, vol. 40, p. 1321–1342, 2016.

[12] Y. Zhang, J. P. Ryan, B. W. Hobson, B. Kieft, A. Romano, B. Barone, C. M. Preston, B. Roman, B.-Y. Raanan, D. Pargett, M. Dugenne, A. E. White, F. H. Freitas, S. Poulos, S. T. Wilson, E. F. DeLong, D. M. Karl, J. M. Birch, J. G. Bellingham, and C. A. Scholin, "A system of coordinated autonomous robots for lagrangian studies of microbes in the oceanic deep chlorophyll maximum," *Science Robotics*, vol. 6, no. 50, p. eabb9138, 2021.

[13] M. Ludvigsen, J. Berge, M. Geoffroy, J. H. Cohen, P. R. D. L. Torre, S. M. Nornes, H. Singh, A. J. Sørensen, M. Daase, and G. Johnsen, "Use of an autonomous surface vehicle reveals small-scale diel vertical migrations of zooplankton and susceptibility to light pollution under low solar irradiance," *Science Advances*, vol. 4, no. 1, p. eaap9887, 2018.

[14] S. Howden, G. Change, B. Currier, C. Grant, J. Jurisich, M. Kim, B. Kirkpatrick, N. Simon, and F. Spada, "Offshore demonstration of an unmanned surface vehicle for autonomous hypoxia monitoring," in *Proceedings of OCEANS*, Mississippi, USA, 2023, pp. 1–5.

[15] S. Yang, J. Huang, X. Xiang, J. Li, and Y. Liu, "Cooperative survey of seabed ROIs using multiple USVs with coverage path planning," *Ocean Engineering*, vol. 268, p. 113308, 2023.

[16] L. Zhao, Y. Bai, and J. K. Paik, "Optimal coverage path planning for USV-assisted coastal bathymetric survey: Models, solutions, and lake trials," *Ocean Engineering*, vol. 296, p. 116921, 2024.

[17] R. Ferretti, M. Bibuli, G. Bruzzone, A. Odetti, S. Aracri, C. Motta, M. Caccia, M. Rovere, A. Mercorella, F. Madricardo, A. Petrizzo, and F. De Pascalis, "Acoustic seafloor mapping using non-standard ASV: technical challenges and innovative solutions," in *Proceedings of OCEANS 2023*, Mississippi, USA, 2023, pp. 1–6.

[18] Maritime Robotics, "World's first uncrewed freight route at sea in the trondheimsfjord," `https://www.maritimerobotics.com/post/world-s-first-uncrewed-freight-route-at-sea-in-the-trondheimsfjord`, 2023, [Online; accessed '1-August-2023].

[19] H. Zheng, W. Xu, D. Ma, and F. Qu, "Dynamic rolling horizon scheduling of waterborne AGVs for inter terminal transportation: Mathematical modeling and heuristic solution," *IEEE Transactions on Intelligent Transportation Systems*, vol. 23, no. 4, pp. 3853 – 3865, 2022.

[20] T. R. Torben, Ø. Smogeli, J. A. Glomsrud, I. B. Utne, and A. J. Sørensen, "Towards contract-based verification for autonomous vessels," *Ocean Engineering*, vol. 270, p. 113685, 2023.

[21] "Maasvlakte Rotterdam, 51.962398° N and 4.056800° E, Google Earth," August 7, 2013, Accessed: 2020-1-9.

[22] J. Liu, X. Yan, C. Liu, A. Fan, and F. Ma, "Developments and applications of green and intelligent inland vessels in china," *Journal of Marine Science and Engineering*, vol. 11, no. 2, p. 318, 2023.

[23] A. Colling, E. van Hassel, R. Hekkenberg, M. Flikkema, W. Nzengu, B. Ramne, E. van der Linden, and N. Kreukniet, "NOVIMAR: Deliverable 1.7: Vessel train handbook," University of Antwerpe, Antwerpe, Belgium, Tech. Rep., 2021.

[24] C. Kent, "Sea train," `https://www.darpa.mil/program/sea-train`, 2020, [Online; accessed '8-September-2023].

CHAPTER 10

Summary and Future Perspectives

10.1 SUMMARY

THE PAST decade has witnessed the rapid development and increasing applications of USVs. However, there are still open challenges in this field to be addressed, and we share the corresponding future perspectives at the end of this book.

This book starts with the ongoing research of USVs over the past decade. Along with the progress of USVs from concept to reality, the USV platforms turn to be more autonomous, versatile and reliable for a variety of scientific, engineering and military applications. The core components on the autonomy of USVs, i.e., the navigation, guidance and control (NGC) methods, have reached a relatively mature phase. Emerging technologies such as the deep learning algorithms still bring about exciting empirical results nonetheless. How to combine the large body of the methodological knowledge with the USV design practices is especially important for the field operational performance. These technologies provide the fundamentals towards the full autonomy of USVs.

Moreover, in this book, we also discuss how these advanced methods can be implemented practically within general USV software and hardware architecture. We note that the implementations and tests in the real marine environment are especially important. This is because methods working theoretically well do not guarantee satisfactory performance in the field due to the complex marine surface conditions. Also, the demands in different applications should be properly considered. Therefore, this book provides the comprehensive knowledge on the USV system.

The overall goal is to reflect on the recent developments and share the authors' research experience on USVs with the community. It is expected to further advance the research towards fully autonomous and finally practically operational USVs to a broader class of maritime applications.

DOI: 10.1201/9781003660590-10

10.2   FUTURE PERSPECTIVES

Further wide applications of the fully autonomous USVs pose a range of particular and entirely new challenges. More technological and practical advances are still needed. We outline several technically open issues relevant with USVs, and give recommendations for future work on these aspects.

10.2.1   Technological perspective

- **The reliable agile maneuverability in complex uncertain marine environments**. Mostly, the existing USV platforms with nominal NGC functionalities can only perform well in smooth weather conditions [1] and low sea states. Therefore, for USVs, especially the small size platforms, the applications have been mainly limited to lakes or inland waterways.

  To tackle the maneuverability issue in uncertain marine environments, there are two ways. The first one is to strengthen the mechanical design. For example, the USV tonnage could be increased as the 120 TEU Yara Birkeland [2]. The USV hull could be constructed in a more stable form as the three-hull Sea Hunter [3] or the quadrotor H20mni-X [4]. Alternatively, by weakening the precise mobility of USVs, the adaptiveness to the environmental forces, e.g., surface wave induced forces, can be improved. The SailDrone [5] and AutoNaut [6] are such representative USVs, which are also called gliders.

  For specific USV designs, the other way to enhance the maneuverability is via the algorithmic adaptiveness and robustness design. Numerous research works have been done. Compared with other transport modes, it is more common to see the installations of millimeter-wave radar on USVs [7]. This is because the USV navigation systems experience the harsh marine weather conditions such as rain, snow, and fog more frequently. The guidance and waypoint generation system of USVs also needs to handle the ubiquitously existing environmental disturbances. Notably, the optimal control and Q-learning based algorithms are applied to consider the ocean currents in [8] and [9], respectively. The precision of the ocean current information is critical in the performance of these algorithms.

  Most of the research works to cope with the uncertain environments lie in the field of control. Conventionally, there are mainly four ways to address the USV internal or external uncertainties:

  - Relying on the system inherent robustness [10]. However, the inherent robustness is limited and can only handle small magnitude of uncertainties.
  - Compensating the disturbances via the feedforward control. The uncertainties in such cases are usually estimated with filters or observers [11, 12] of which the convergence speed should be strictly guaranteed.
  - Designing robustness mechanisms, e.g., using H_∞ [13] or LQG [14]. However, most of this class of works assume specific features of the

uncertainties, e.g., stochastic with specific distributions [15, 16] or bounded with known boundaries [17].

– Designing adaptiveness mechanisms so that the system uncertain parameters could evolve with the dynamic environment or the system states [18, 19]. Mostly, the adaptiveness mechanism also relies on the prior known information of the uncertainties.

Overall, it is still challenging for the conventional robust control methods to deal with the complex marine uncertainties effectively. Most of the existing research works are still limited to simulations or ideal weather conditions. Novel methods to represent and deal with the maritime uncertainties for USVs should be developed.

The emerging learning-based technologies are featured with the ability to approximate any nonlinear dynamics and the flexibility in handling time varying systems. Thus, the learning-based approaches have been explored extensively for USV modeling and control problems recently [20, 21, 22, 23, 24]. The training data is crucial in applying the learning-based approaches. Moreover, it is in general difficult to provide provable control performance guarantee which is especially important in safety critical situations. Nonetheless, the learning-based methodologies still show great potential for USVs against uncertainties. Hopefully, the reliable agile maneuverability of USVs could benefit from the advances in the systematic analysis and synthesis of the machine learning and deep learning algorithms.

- **Cooperative swarm intelligence**. Another important direction is the cooperative swarm intelligence. The swarm can be homogeneous or heterogeneous. Broadly, homogeneous USVs in swarms achieve much higher efficiency and can be more robust to local failures [25, 26, 27]. Thus, swarms of USVs have been widely used in military, cooperative target search [28] and circumnavigation [1], cooperative waterborne transportation [29, 30, 31, 32], and cooperative bathymetry [33, 34], etc.

Due to the cross-medium feature, USVs can also work cooperatively with heterogeneous platforms such as unmanned ariel vehicles (UAVs) and AUVs. As discussed before, USVs can collaborate with AUVs for a variety of oceanographic sampling and mapping tasks [35, 36, 37, 38]. In most of these tasks, USVs and AUVs show complementary advantages. USVs have access to the communication and localization resources, and AUVs can operate in the deep sea. In the maritime environment, UAVs can also work cooperatively with USVs [39][40] since UAVs usually have larger coverage fields than USVs. To fully explore the advantages of multiple types of unmanned vehicles, a joint design of heterogeneous UAV-USV-UUV, i.e., 3U, swarm network is proposed in [41] for cooperative target hunting. Simulation results show that the target hunting success rate is improved significantly with the 3U network.

However, it should be noted that the USV swarms have potential technological challenges. Firstly, unlike other types of unmanned swarms, the communication in the maritime environment is costly and not reliable. Therefore, the information sharing which is essential for the cooperation between different USV agents is quite difficult. Even if the communication is available, it usually experiences delays and packet losses, which complicate the cooperative problem and deteriorate the swarm performance significantly. With an appropriate communication scheme, the control structure of the swarm can be designed to be either centralized or distributed. Secondly, also different with other types of unmanned swarms, USVs maneuver in the fluid where the hydrodynamic interactions between USVs in the proximity cannot be neglected. Therefore, hydrodynamic interactions could give rise to the couplings also in the USV dynamics. The swarm's operational performance could be improved if the hydrodynamic interactions are properly considered.

10.2.2 Practical perspective

- **Computational efficiency.** For USVs maneuvering fully autonomously in the complex and dynamic environment, it is important that the USVs could make decisions timely. In this way, the system behaviors could be adjusted according to, either the dynamic tasks or the dynamic environment, to maintain the system performance and safety. Dynamic decision makings require that the on board computations are performed efficiently. On the one hand, more advanced computational hardware can be used. On the other hand, the state-of-the-art NGC algorithms could be deployed in a hierarchical, modular, and distributed way to achieve more tractable solutions. For cases where timely update and feedback are critical, a reliable decision recovery mechanism, e.g., making safe but not necessarily optimal decisions based on the previous solution, could be necessary.

- **Supporting infrastructures.** USVs show many differences with the conventional ships. The most prominent difference is the intelligence. Besides the on board intelligence with the USV NGC system, the shore supporting infrastructures are also important. For example, the communication network and protocols for the USV-to-USV and USV-to-shore are important in realizing the coordination of USVs. The shore-side perception facilities could also provide important environment information for the USVs especially in the inland waterways. For USVs, special deployment and recovery facilities are also important for the efficient and safe operation of USVs.

- **Cyber security and privacy.** Cyber security risks vary depending on levels of autonomy. The risks might exist in the USV automation system itself or in the network via which the USV communicates with the outside nodes. Currently, the security and privacy issues of USVs have not received enough attention. However, like other autonomous systems, USVs are prone to various attacks which might cause heavy damages especially in hostile situations.

- **Industrial awareness and cost-benefit analysis.** In the process to become a type of mature product, it is essential to analyze from the commercial perspective. Firstly, more work should be done to raise the awareness of USVs in relevant industries. For example, USVs show great potential in the autonomous waterborne transport, and more pilot or experimental platforms should be built for tests and demonstrations [42]. Secondly, research on a systematic cost-benefit analysis of applying USVs in different fields could be carried out. Data in terms of labor, time, and cost savings, etc., could be collected for the cost-benefit analysis. Moreover, analytical and simulation models with USVs being involved could be built to predict and analyze the potential strengths and weaknesses of such systems [43, 44].

- **Legal perspective.** USVs considered in this book are designed to be cooperative to optimize an overall cost and no priorities are assigned. However, in practice, USVs could belong to different terminals and operated by different companies. A more complicated scenario involves manned marine vehicles which cannot be controlled to be cooperative. In these cases, non-cooperative coordination approaches should be investigated. One option is to assign priorities to vehicles according to certain rules, e.g., COLREGs [45] as elaborated in Chapter 3. Flexibility and optimality could be the concerns for the rule-based coordination. More work could be done on how to integrate rule-based coordination into the optimization-based coordination strategies to handle mixed traffic situations. Moreover, unmanned vehicles also involve the responsibility issues among the producer, owner and user especially when accidents happen. Laws should be established to handle such cases.

Bibliography

[1] Z. Yan, H. Zheng, Z. Jiang, and W. Xu, "Distributed control of unmanned marine vehicles for target circumnavigation in communication-denied environments," *IEEE/ASME Transactions on Mechatronics*, early access, 2024.

[2] T. K. Grose, "Skip the skipper," *ASEE Prism*, vol. 31, no. 5, pp. 8–8, 2022.

[3] K. J. Casola, "System architecture and operational analysis of medium displacement unmanned surface vehicle sea hunter as a surface warfare component of distributed lethality," Ph.D. dissertation, Naval Postgraduate School, Monterey, USA, 2017.

[4] N. Miskovic, D. Nad, and I. Rendulic, "Tracking divers: An autonomous marine surface vehicle to increase diver safety," *IEEE Robotics & Automation Magazine*, vol. 22, no. 3, pp. 72–84, 2015.

[5] C. Gentemann, J. P. Scott, P. L. Mazzini, C. Pianca, S. Akella, P. J. Minnett, P. Cornillon, B. Fox-Kemper, I. Cetinić, T. M. Chin *et al.*, "Saildrone: Adaptively sampling the marine environment," *Bulletin of the American Meteorological Society*, vol. 101, no. 6, pp. E744–E762, 2020.

[6] A. Dallolio, H. B. Bjerck, H. A. Urke, and J. A. Alfredsen, "A persistent sea-going platform for robotic fish telemetry using a wave-propelled USV: Technical solution and proof-of-concept," *Frontiers in Marine Science*, vol. 9, p. 857623, 2022.

[7] Z. Wang and Y. Zhang, "Estimation of ship berthing parameters based on Multi-LiDAR and MMW radar data fusion," *Ocean Engineering*, vol. 266, p. 113155, 2022.

[8] D. Ma, S. Hao, W. Ma, H. Zheng, and X. Xu, "An optimal control-based path planning method for unmanned surface vehicles in complex environments," *Ocean Engineering*, vol. 245, p. 110532, 2022.

[9] Y. Yu, H. Zheng, and W. Xu, "Learning and sampling-based informative path planning for AUVs in ocean current fields," *IEEE Transactions on Systems, Man, and Cybernetics: Systems*, vol. 55, no. 1, pp. 51–62, 2025.

[10] D. A. Allan, C. N. Bates, M. J. Risbeck, and J. B. Rawlings, "On the inherent robustness of optimal and suboptimal nonlinear MPC," *Systems & Control Letters*, vol. 106, pp. 68–78, 2017.

[11] H. S. Halvorsen, H. Øveraas, O. Landstad, V. Smines, T. I. Fossen, and T. A. Johansen, "Wave motion compensation in dynamic positioning of small autonomous vessels," *Journal of Marine Science and Technology*, vol. 26, pp. 693–712, 2021.

[12] G. X. Wu, Y. Ding, T. Tahsin, and A. Incecik, "Adaptive neural network and extended state observer-based non-singular terminal sliding mode tracking control for an underactuated USV with unknown uncertainties," *Applied Ocean Research*, vol. 135, p. 103560, 2023.

[13] Y. Huang and P. Huang, "An improved bounded real lemma for h∞ robust heading control of unmanned surface vehicle," *Proceedings of the Institution of Mechanical Engineers, Part I: Journal of Systems and Control Engineering*, vol. 237, no. 10, pp. 1812–1821, 2023.

[14] T. Asfihani, D. K. Arif, F. P. Putra, M. A. Firmansyah *et al.*, "Comparison of LQG and adaptive PID controller for ASV heading control," in *Journal of Physics: Conference Series*, vol. 1218, no. 1, 2019, p. 012058.

[15] H. Zheng, R. R. Negenborn, and G. Lodewijks, "Explicit use of probabilistic distributions in robust predictive control of waterborne AGVs - a cost-effective approach," in *Proceedings of the 15th European Control Conference*, Aalborg, Denmark, 2016, pp. 1278–1283.

[16] ——, "Robust distributed predictive control of waterborne AGVs—A cooperative and cost-effective approach," *IEEE Transactions on Cybernetics*, vol. 48, no. 8, pp. 2449–2461, 2018.

[17] H. Zheng, J. Wu, W. Wu, and Y. Zhang, "Robust dynamic positioning of autonomous surface vessels with tube-based model predictive control," *Ocean Engineering*, vol. 199, p. 106820, 2020.

[18] R. Skjetne, T. I. Fossen, and P. V. Kokotović, "Adaptive maneuvering with experiments for a model ship in a marine control laboratory," *Automatica*, vol. 41, no. 2, pp. 289–298, 2005.

[19] L. Chen, S.-L. Dai, and C. Dong, "Adaptive optimal tracking control of an underactuated surface vessel using Actor–Critic reinforcement learning," *IEEE Transactions on Neural Networks and Learning Systems*, vol. 35, no. 6, pp. 7520 – 7533, 2024.

[20] H. Zheng, J. Li, Z. Tian, C. Liu, and W. Wu, "Hybrid physics-learning model based predictive control for trajectory tracking of unmanned surface vehicles," *IEEE Transactions on Intelligent Transportation Systems*, vol. 25, no. 9, pp. 11522 – 11533, 2024.

[21] W. Yuan and X. Rui, "Deep reinforcement learning-based controller for dynamic positioning of an unmanned surface vehicle," *Computers and Electrical Engineering*, vol. 110, p. 108858, 2023.

[22] Y. Zhang, S. Li, and J. Weng, "Learning and near-optimal control of underactuated surface vessels with periodic disturbances," *IEEE Transactions on Cybernetics*, vol. 52, no. 8, pp. 7453–7463, 2021.

[23] D. Chwa, "Adaptive neural output feedback tracking control of underactuated ships against uncertainties in kinematics and system matrices," *IEEE Journal of Oceanic Engineering*, vol. 46, no. 3, pp. 720–735, 2020.

[24] N. Wang, Y. Gao, H. Zhao, and C. K. Ahn, "Reinforcement learning-based optimal tracking control of an unknown unmanned surface vehicle," *IEEE Transactions on Neural Networks and Learning Systems*, vol. 32, no. 7, pp. 3034–3045, 2020.

[25] Z. Wang, S. Yang, X. Xiang, A. Vasilijević, N. Mišković, and Đ. Nađ, "Cloud-based mission control of USV fleet: Architecture, implementation and experiments," *Control Engineering Practice*, vol. 106, p. 104657, 2021.

[26] C. Pan, Z. Peng, Y. Li, B. Han, and D. Wang, "Flocking of under-actuated unmanned surface vehicles via deep reinforcement learning and model predictive path integral control," *IEEE Transactions on Instrumentation and Measurement*, 2024.

[27] Y. He, L. Chen, J. Mou, Q. Zeng, Y. Huang, P. Chen, and S. Zhang, "Ship emission reduction via energy-saving formation," *IEEE Transactions on Intelligent Transportation Systems*, vol. 25, no. 3, pp. 2599–2614, 2024.

[28] G. Zhang, S. Liu, X. Zhang, and W. Zhang, "Event-triggered cooperative formation control for autonomous surface vehicles under the maritime search operation," *IEEE Transactions on Intelligent Transportation Systems*, vol. 23, no. 11, pp. 21 392–21 404, 2022.

[29] H. Zheng, R. R. Negenborn, and G. Lodewijks, "Fast ADMM for distributed model predictive control of cooperative waterborne AGVs," *IEEE Transactions on Control Systems Technology*, vol. 25, no. 4, pp. 1406–1413, 2016.

[30] ——, "Closed-loop scheduling and control of waterborne AGVs for energy-efficient inter terminal transport," *Transportation Research Part E: Logistics and Transportation Review*, vol. 105, pp. 261– 278, 2017.

[31] L. Chen, Y. Huang, H. Zheng, H. Hopman, and R. Negenborn, "Cooperative multi-vessel systems in urban waterway networks," *IEEE Transactions on Intelligent Transportation Systems*, vol. 21, no. 8, pp. 3294–3307, 2019.

[32] C. Liu, J. Qi, X. Chu, M. Zheng, and W. He, "Cooperative ship formation system and control methods in the ship lock waterway," *Ocean Engineering*, vol. 226, p. 108826, 2021.

[33] S. Yang, J. Huang, X. Xiang, J. Li, and Y. Liu, "Cooperative survey of seabed ROIs using multiple USVs with coverage path planning," *Ocean Engineering*, vol. 268, p. 113308, 2023.

[34] L. Zhao, Y. Bai, and J. K. Paik, "Optimal coverage path planning for USV-assisted coastal bathymetric survey: Models, solutions, and lake trials," *Ocean Engineering*, vol. 296, p. 116921, 2024.

[35] K. Zwolak, B. Simpson, B. Anderson, E. Bazhenova, R. Falconer, T. Kearns, H. Minami, J. Roperez, A. Rosedee, H. Sade, N. Tinmouth, R. Wigley, and Y. Zarayskaya, "An unmanned seafloor mapping system: The concept of an AUV integrated with the newly designed USV SEA-KIT," in *In Proceedings of OCEANS*, Aberdeen, UK, 2017, pp. 1–6.

[36] A. A. Proctor, Y. Zarayskaya, E. Bazhenova, M. Sumiyoshi, R. Wigley, J. Roperez, K. Zwolak, S. Sattiabaruth, H. Sade, N. Tinmouth, B. Simpson, and S. M. Kristoffersen, "Unlocking the power of combined autonomous operations with underwater and surface vehicles: Success with a deep-water survey AUV and USV mothership," in *Proceedings of OCEANS*, Kobe, USA, 2018, pp. 1–8.

[37] A. Guerrero-González, F. García-Córdova, and F. Ortiz, "A multirobot platform based on autonomous surface and underwater vehicles with bio-inspired neuro-controllers for long-term oil spills monitoring," *Autonomous Robots*, vol. 40, p. 1321–1342, 2016.

[38] C. R. German, M. V. Jakuba, J. C. Kinsey, J. Partan, S. Suman, A. Belani, and D. R. Yoerger, "A long term vision for long-range ship-free deep ocean

operations: Persistent presence through coordination of autonomous surface vehicles and autonomous underwater vehicles," in *Proceedings of IEEE/OES Autonomous Underwater Vehicles (AUV)*, Southampton, UK, 2012, pp. 1–7.

[39] Y. Zhu, S. Li, G. Guo, P. Yuan, and J. Bai, "Formation control of UAV–USV based on distributed event-triggered adaptive MPC with virtual trajectory restriction," *Ocean Engineering*, vol. 294, p. 116850, 2024.

[40] J. Li, G. Zhang, Q. Shan, and W. Zhang, "A novel cooperative design for USV–UAV systems: 3-D mapping guidance and adaptive fuzzy control," *IEEE Transactions on Control of Network Systems*, vol. 10, no. 2, pp. 564–574, 2023.

[41] W. Wei, J. Wang, Z. Fang, J. Chen, Y. Ren, and Y. Dong, "3U: Joint design of UAV-USV-UUV networks for cooperative target hunting," *IEEE Transactions on Vehicular Technology*, vol. 72, no. 3, pp. 4085–4090, 2023.

[42] Maritime Robotics, "World's first uncrewed freight route at sea in the trondheimsfjord," https://www.maritimerobotics.com/post/world-s-first-uncrewed-freight-route-at-sea-in-the-trondheimsfjord, 2023, [Online; accessed '1-August-2023].

[43] J. L. Dantas and G. Theotokatos, "A framework for the economic-environmental feasibility assessment of short-sea shipping autonomous vessels," *Ocean Engineering*, vol. 279, p. 114420, 2023.

[44] H. Nordahl, D. A. Nesheim, and E. Lindstad, "Autonomous ship concept evaluation–quantification of competitiveness and societal impact," in *Journal of Physics: Conference Series*, vol. 2311, no. 1. IOP Publishing, 2022, p. 012020.

[45] H. Zheng, J. Zhu, C. Liu, H. Dai, and Y. Huang, "Regulation aware dynamic path planning for intelligent ships with uncertain velocity obstacles," *Ocean Engineering*, vol. 278, p. 114401, 2023.

Appendix

A.1 RELEVANT COLREGS RULES WITH CHAPTER 3

- RULE 13: Overtaking. Any vessel overtaking any other shall keep out of the way of the vessel being Overtaken.

- RULE 14: Head-on Situation. When two power-driven vessels are meeting on reciprocal or nearly reciprocal courses so as to involve risk of collision each shall alter her course to starboard so that each shall pass on the port side of the other.

- RULE 15: Crossing Situation. When two power-driven vessels are crossing so as to involve risk of collision, the vessel which has the other on her own starboard side shall keep out of the way and shall, if the circumstances of the case admit, avoid crossing ahead of the other vessel.

- RULE 16: Action by Give-way Vessel. Every vessel which is directed by these rules to keep out of the way of another vessel shall, so far as possible, take early and substantial action to keep well clear.

- RULE 17: Action by Stand-on Vessel. Where by any of these rules one of two vessels is to keep out of the way the other shall keep her course and speed.

DOI: 10.1201/9781003660590-A

A.2 DYNAMICS PARAMETERS USED IN CHAPTER 5

In the appendix, we present the specific parameters that are involved in (5.2) in Chapter 5 as follows:

$$a_{11} = \frac{m_{22}}{m_{11}}, \ a_{12} = \frac{m_{23}}{m_{11}}, \ a_{13} = \frac{X_u}{m_{11}}, \ a_{14} = \frac{1}{m_{11}}, \ a_{21} = -\frac{m_{11}m_{23} - m_{22}m_{23}}{m_{22}m_{33} - m_{23}m_{32}},$$

$$a_{22} = \frac{m_{23}m_{23} - m_{11}m_{33}}{m_{22}m_{33} - m_{23}m_{32}}, \ a_{23} = \frac{m_{33}Y_v - m_{23}N_v}{m_{22}m_{33} - m_{23}m_{32}}, \ a_{24} = \frac{m_{33}Y_r - m_{23}N_r}{m_{22}m_{33} - m_{23}m_{32}},$$

$$a_{25} = \frac{m_{33}}{m_{22}m_{33} - m_{23}m_{32}}, \ a_{26} = \frac{-m_{23}}{m_{22}m_{33} - m_{23}m_{32}}, \ a_{31} = -\frac{m_{22}m_{22} + m_{11}m_{22}}{m_{22}m_{33} - m_{23}m_{32}},$$

$$a_{32} = \frac{m_{11}m_{32} - m_{22}m_{23}}{m_{22}m_{33} - m_{23}m_{32}}, \ a_{33} = \frac{m_{22}N_v - m_{32}Y_v}{m_{22}m_{33} - m_{23}m_{32}}, \ a_{34} = \frac{m_{22}N_r - m_{32}Y_r}{m_{22}m_{33} - m_{23}m_{32}},$$

$$a_{35} = \frac{-m_{32}}{m_{22}m_{33} - m_{23}m_{32}}, \ a_{36} = \frac{m_{22}}{m_{22}m_{33} - m_{23}m_{32}}, \ d_1 = \frac{X_{|u|u}|u| + X_{uuu}u^2}{m_{11}}u + d_x,$$

$$d_2 = \frac{m_{33}(Y_{|v|v}|v| + Y_{|r|v}|r|) - m_{23}(N_{|v|v}|v| + N_{|r|v}|r|)}{m_{22}m_{33} - m_{23}m_{32}}v$$

and

$$+ \frac{m_{33}(Y_{|v|r}|v| + Y_{|r|r}|r|) - m_{23}(N_{|v|r}|v| + N_{|r|r}|r|)}{m_{22}m_{33} - m_{23}m_{32}}r + d_y$$

$$d_3 = \frac{m_{22}(N_{|v|v}|v| - N_{|r|v}|r|) - m_{32}(Y_{|v|v}|v| + Y_{|r|v}|r|)}{m_{22}m_{33} - m_{23}m_{32}}v$$

$$+ \frac{m_{22}(N_{|v|r}|v + N_{|r|r}|r|) - m_{32}(Y_{|v|r}|v| + Y_{|r|r}|r|)}{m_{22}m_{33} - m_{23}m_{32}}r,$$

where, $m_{11} = m - X_{\dot{u}}$, $m_{22} = m - Y_{\dot{v}}$, $m_{23} = mx_G - Y_{\dot{r}}$, $m_{32} = mx_G - N_{\dot{v}}$, $m_{33} = I_z - N_{\dot{r}}$. m is the mass of the ASV; I_z is the moment of inertia in the yaw direction; and x_G is the distance between the center of gravity of the ASV to the center of o_b. m_{11}, m_{22}, m_{23}, m_{32} and m_{33} are the components of the inertia matrix M

$$M = \begin{bmatrix} m_{11} & 0 & 0 \\ 0 & m_{22} & m_{23} \\ 0 & m_{32} & m_{33} \end{bmatrix}.$$

Index

Note: – Pages in *Italics* refer to figures and pages in **bold** refer to tables.

For Product Safety Concerns and Information please contact our EU
representative GPSR@taylorandfrancis.com
Taylor & Francis Verlag GmbH, Kaufingerstraße 24, 80331 München, Germany

www.ingramcontent.com/pod-product-compliance
Lightning Source LLC
Chambersburg PA
CBHW082006190326
41458CB00010B/3097

9 781041 115595